THE
HAPPY
DOG
OWNER

THE
HAPPY
DOG
OWNER

Dr CARRI WESTGARTH

WELBECK

Published by Welbeck
An imprint of Welbeck Non-Fiction Limited,
part of Welbeck Publishing Group.
20 Mortimer Street,
London W1T 3JW

First published by Welbeck in 2021

A CIP catalogue record for this book is available from the British Library

ISBN
Paperback – 9781787396555

Typeset by seagulls.net
Printed at CPI UK

2 4 6 8 10 9 7 5 3 1

www.welbeckpublishing.com

CONTENTS

CHAPTER 1

INTRODUCTION

Few people are lucky enough to live out their lives without ever having to navigate their way through some form of illness, be that physical or mental. Even when we're not ill, the pressure to take more exercise and adopt a healthier lifestyle can often seem relentless and, dare I say it, a bit short on joy. This is ironic given that the other authentically human pursuit is our eternal quest to be happy – to free ourselves of stress, worries and malaise. But what if there was a really simple way to help many people live a happier, healthier life? A way that doesn't involve popping pills, or surgery, or expensive re-designs of the places we live and work? A way that involves something many of us already have in our homes, a some-thing that at this very moment is probably waiting patiently with a wagging tail for their owner to appear. Of course, I am talking about … a dog.

As many other dog owners will tell you, I can't imagine my life without dogs. Actually, I can: it would be miserable and I would weigh a lot more. My dogs make every day at least a little bit better. When they jump on my bed in the morning and climb on to my chest until I put my phone down and rub their ears instead, I start the day calmly and with

ROXIE THE PUG-CHIHUAHUA

a smile. When I come home in the evening, they are always pleased to see me, which is a welcome antidote to a hard day's work. It fills my heart with joy when they snuggle under a blanket watching TV with my seven-year-old son. And most important of all, every single day, come rain or shine, they make me take them out for a walk. For me, owning dogs is a 99 per cent positive experience, but this situation didn't just miraculously happen – we had to work at it.

Life can be particularly hard for some people. For me, while anxiety and depression have both occasionally been an issue, my big health challenge is managing chronic back pain. In one painful episode, I struggled my way to the bathroom and was sat on the toilet when I accidentally dropped the paper roll. I could only watch, devastated, as it slowly rolled away from me across the floor before coming to a standstill under the sink, over a metre away from me. After feeling sorry for myself for a minute or so because I couldn't bear the pain of trying to get to it, an idea occurred. "Roxie!" I called. A little brown head with big eyes popped around the (thankfully not locked) door – my pug-chihuahua mix. "Hey Roxie, fetch it!" I said, and pointed. Off she trotted, pounced on the paper roll, grabbed it with her mouth, brought it over and dropped it at my feet. Good girl! Crisis averted! Clearly, that day the presence of my dog was of huge benefit to dealing with my health symptoms. She also got to have a bit of fun while helping me.

In both my professional and personal life, I have experienced countless examples of dogs helping to improve their owners' quality of life even beyond what might be expected of the family pet. The following two stories immediately spring to mind:

"My condition is neurocardiogenic syncope. I get light-headed, experience brain fog and chronic fatigue, which all get worse the longer I'm upright for, or when exercising, showering, getting dressed, sitting down, or driving. Eventually, I'll pass out. My dog will let me know if I'm just feeling light-headed or if I'm actually going to pass out. Her alerts are natural, she just started doing it. She gives me a worried look or puts her head on my knee." – **Nadine**

"I have severe PTSD as well as high social anxiety, fibromyalgia and depression. My dog is the reason I get up each morning. When I am going through an anxiety attack he will move into me and lick me, helping me focus and calm quickly. He also knows if I'm having flashbacks or sensory overload as he will bring me something to play with, or if I'm sat, curl up on me to soothe me. When he senses something, his posture changes and he will touch me with his nose without the command, or he makes a noise and gets me to focus on him. It's not a bark but a little yip. A lot of what he does is emotional therapy rather than physical aid, though he enables me to leave the house if I'm really struggling." – **Rachel**

While these two examples may not be directly relevant to everyone's needs, I wanted to give you an idea of what's possible, and show how your dog can help, whatever your situation. Some dogs, through training or natural responses, can go even further in assisting with their owners' health needs, a prime example being official assistance (UK) or service (USA) dogs. But whatever the level of assistance required, the average dog owner might struggle to know how best to train for these behaviours. The good news is that the process can be broken down into simple, achievable steps. Once you understand the principles, there is no reason why you can't train your dog to help with all sorts of everyday tasks, from being a canine alarm clock to simply learning to cuddle on command. In addition, there are small adjustments that can be made to our daily interactions with our pets that can improve our health immensely. This book will show you how. Once you have read this book, you will be able to analyze your relationship with your dog, and use your new-found practical training skills and understanding of your dog to assist your well-being needs.

The advice in this book isn't just based on my personal (and some might argue, lucky) experiences. I have spent the last 15 years conducting scientific research into how people interact with their dogs, and the positive and negative impacts on *their* health that occur as a result. I am part of a global network of academic researchers working collectively to better understand how pets impact our lives – and also how us owning them affects their welfare. I have also spent many more years helping hundreds of dogs and their owners: matching prospective owners with homeless dogs and new puppies; teaching obedience and tricks; counselling owners through dealing with problematic behaviours such as aggression; and training official assistance dogs to work with handlers with disabilities such as hearing loss. These real-life, practical experiences have informed my research questions, and my findings impact the advice I give to people about how to source and train a well-adjusted family pet. I am also called upon to advise organizations and charities developing policies about managing dog walkers and other aspects of owning pets in today's society. With this book, I hope to both impart what my colleagues and I know about forging robust relationships that foster good animal welfare and owner well-being, and show you how to train your dog to train YOU to be happier and healthier.

One in four households in the UK owns a dog, and this figure is even higher in the USA and Australia. Many dog owners claim that their health and well-being have been improved by pet ownership, and some doctors even recommend dogs to their patients. As I will expand on later, many of the health benefits start with the simple dog walk. Once, when I was being interviewed on the radio, the presenter suggested it was "exercise by stealth", a phrase that I love. Just because you're not running on a boring treadmill in a gym doesn't mean it's not good for you! Physical activity is important for preventing obesity, cancer and depression, among other nasties. Research interviews I have conducted with dog owners reveal that dogs provide a motivation to get out and walk that is a far more effective stimulus for increasing physical activity than typical public-health campaigns. Dog walking is also a highly social activity – people are much more likely to engage in a friendly chat if you've got a dog with you. And there are all sorts of other social benefits too, some of which you might not expect.

In my early twenties, I adopted Jasmyn, a black crossbreed (according to DNA analysis, mostly cocker and springer spaniel, with a great-grandparent flat-coated retriever thrown into the mix) from the rescue centre where I worked at the time. A few years later, I also inherited Ben, a dopey tri-coloured border collie, from my parents.

JASMYN THE CROSS-BREED

Ben was a farming reject, who was slowly rehabilitated to normal pet life with my dad and stepmum for a few years, alongside their other collie, Sky (whom we will meet later). In his latter years, Ben came to live with me, to provide some company for Jas.

One frosty early morning, Jas, Ben and I were out on our usual walk, coming home from the woods. I was in a bubble of my own thoughts when I noticed that the dogs, Ben in particular, were acting strangely, softly whining and edging into the road. Initially confused and a bit annoyed by them, I decided to follow their lead after they persisted in this behaviour. They took me across the road to a house where, it turned out, they had found an injured elderly woman who had fallen over, out of sight, on her driveway. I was able to help her inside and get medical attention. Without my dogs, who knows how long it would have been before someone helped her? This personal experience supports the scientific evidence that our dogs not only impact our own lives, but also the people around us.

BEN THE COLLIE

Dog owners often report that their animals relieve stress and help to reduce feelings of loneliness – no mean feat in a world where so many of us are increasingly disconnected from meaningful social interaction, conducting much of our work or social life online and through screens. This became even more apparent during the coronavirus pandemic "lockdowns" (during which I am writing this book) and our pets proved to be a great source of emotional support throughout this unprecedented time.

However, before we skip off into the sunset with our very own Lassie, it is vital that we understand this: the mere fact of dog ownership is not enough to transform your and your dog's health and happiness. If the media are to be believed, owning a dog is the perfect antidote to the recent pandemic or the ongoing epidemics of obesity, heart disease and depression. Surprisingly, though, science has so far failed to find strong evidence of a causal relationship. So why are these benefits not universal? Fascinatingly, while some studies do show that dogs have positive effects on our mental and physical well-being, others show no difference at all, and still others even point to relatively poorer health among dog owners. Deeper analysis shows that part of the problem seems to result from the extraordinary fact that many dog owners don't walk their dogs at all. Furthermore, it seems that not all owners always find their dogs so wonderful to be around.

When you think about it a little more, perhaps this is unsurprising. Living with a dog can be relaxing at one moment (walking in the park with friends) and stressful the next (when the dog runs off and barks at a stranger). Many of my friends' full-time careers are centred around providing support and guidance to owners whose pets are causing them serious stress due to the display of problematic behaviours. These may be normal for the dog but undesirable to the owners, or a serious concern for the welfare of both. When things go wrong and there are behavioural issues, dog ownership can be frustrating and upsetting. The impact of truly serious problems – such as when a dog displays aggressive behaviour – can be life-changing for both the victim and the dog. Other colleagues have to deal with the fallout when things get so bad that pups that were once much-loved end up being given up for adoption or taken to be put down.

Part of the challenge is that what we want, or even need, from our dogs has changed immensely over recent years. Dogs occupy a very different place in our hearts and our houses than they did just a few decades ago. Up until the latter half of the last century, dogs played a primarily utilitarian role in our lives. Most were bred and owned for a specific function, such as helping us hunt or being a hot water bottle, with few expectations beyond that.

Human–animal interaction researchers Dr Rebekah Fox and Professor Nancy Gee interviewed 20 pet owners and 21 animal professionals from the UK about their perceptions of changes in pet-keeping practices over the past 30 to 40 years and how their own pet-keeping practices had changed over time, or differed from those of their parents or relatives.[1] While the current situation may seem "normal" to us right now, in fact, Rebekah's research shows that animals have become simultaneously subject to increasing control of their behaviour and movements in public spaces as well as increasingly integrated into the human home and family. The now common culture of "responsible" pet ownership drives our decisions as owners and also our surveillance and judgement of others. Our status in society is increasingly defined by what dog we own and how we are viewed to look after it and protect others from it. Our view of dogs has changed not only socially but also legally, including prescriptions about how our dogs are allowed to behave, how to identify who owns them and how we should dispose of their faeces.[2] Indeed, in a mortifying moment, I was once fined £50 when Jasmyn did a "second poop" while my back was turned as I was practising a "stay" command with Ben and Roxie.

Erica Peachey was one of the very first people to call herself a pet behaviour counsellor, when in 1990 she began running novel puppy-training classes, and also helping owners with their pets' problems. By the time I wrote to her requesting work experience at the beginning of my PhD studies in 2005, having conveniently just moved to a house a couple of streets away from her, she had been practicing for 15 years already. Now, we have doubled that with a combined achievement of 45 years working with dog owners, 15 of these together, and often reflect on the changes we have observed.

Erica tells me:

"While I have been practising, the way we live and work has changed and our expectations of our dogs have also changed massively. Dogs have become far more humanized and in some ways given greater consideration. However, this can mean, they are less allowed to be dogs. The good news is that there appears to be much more deliberate decision-making around getting a dog – for example, fewer dogs being acquired because a friend's dog had an accidental litter. People who traditionally wouldn't have had dogs – because they are allergic and have never really been around dogs – now have the ability to get one and add to their family. They often select a designer cross-breed that they think doesn't moult (but they sometimes do). Some dogs may struggle to cope with periods of intense, and often highly physical, inter-action with us, interspersed between long periods of being left to their own devices while our lives are focused on screens relating to work or entertainment. At the same time, the world has become a much less black-and-white place, and the societal rules about what we should and shouldn't do regarding our dogs less clear. There is so much information available to us, but it's from the contrasting viewpoints of different experts who each sound so plausible and likable, so it is hard to distinguish between what to believe and what to ignore."

In today's society, we expect our dog to be our best friend, childminder, dress-up doll, entertainment, protector and wingman. We expect them to be always by our side – except when we are out at work for 12 hours a day – and to be entirely tolerant of everything modern life can throw at them. Given that so many of us are struggling to cope, it's no wonder our dogs are too. With all these developments, it is clear that we have not just a moral responsibility but also a legal one to rethink what is happening. We must invest time and effort into training our dogs, and, inasmuch as we can, prevent issues from occurring. We must also distinguish the hype from the evidence, all of which this book intends to achieve.

It is time to re-evaluate what we want from our dogs, and what they need from us. To me, it seems clear that only a certain kind of dog ownership really benefits the owner's health. This can only happen if the dog is trained correctly and the owner does certain activities with it. To achieve this, we need to make intentional choices around where we get our dogs from, how we train them, and what activities we do with them. You can only get out what you put in. In simple terms, only a happy, obedient dog leads to a relaxed, healthy owner.

My mission in *The Happy Dog Owner* is to teach the reader how to maximize the benefits of dog ownership and minimize the risks. Completing some simple training tasks is essential, but the training must be based on mutual trust, respect and caring. It must use kind methods and create the right kind of relationship between an owner and their dog. Although our dogs might already be enhancing our well-being, there is always more we can do to improve our relationship with them, and so increase the benefits they can bring. In an ideal scenario, a synergistic relationship develops, in which the well-being of both dog and owner is optimized, resulting in improved levels of happiness and health for both parties. What's not to like?

This book does not intend to be a thorough examination of all we know about dogs – their cognition, biology, welfare or history. There are many fantastic books out there that cover these areas (see the section on further reading). Neither can it be an exhaustive review of *all* the scientific knowledge we have about pets and human health. However, it will guide you through the main principles, making it clear what we know (and still don't know) about our relationships with dogs and how owning them affects our physical and mental health. Along the way, we will meet experts making inroads in these areas, and pick their brains about how to translate their knowledge into everyday practices we can use with our own pets. Through this exploration, we can become happier owners of happier dogs.

CHAPTER 2

EXAMINING "THE PET EFFECT"

If you Google "pets and health", the search results may look something like this: "Healthy pets – healthy people!"; "Why having a pet is good for your health"; "The Power of Pets"; "Mood-boosting Power of Pets"; "10 Benefits of Owning a Pet"; "27 Ways Pets Improve Your Health", and so on. If you change the search terms to "dogs and health", you get a host of similar headlines. My work here is done – that was an easy book to write! If only it were that simple. We need to take a closer look at the theoretical support and scientific evidence behind those fluffy headlines. This is critical to manage our expectations of what our pets can do for us and to fully comprehend the ways in which owning a dog is likely to positively impact our well-being.

Dogs are a common sight in everyday life. You would struggle to find anyone who didn't have some form of contact with dogs, even through a family member or friend, if they didn't own one themselves. Yet we know surprisingly little about the implications of being around dogs. This book is concerned with the relationships we have with domestic pet dogs, but there are arguably many more dogs on the planet that live out their lives in strikingly different circumstances, whether feral, stray, or community/village owned. We will not be discussing those dogs in this book, but instead focusing on the pet dogs who live – without any clear "function", at first glance – in our homes. There are also ethical and academic debates about whether it is appropriate to talk about "owning" dogs as if they are property – like you own a house or a pair of shoes. Some prefer the term "pet guardian" or "pet parent", but like many of my colleagues, I prefer to stick with the word "ownership" because that has historically been, and remains, common usage.[1] Furthermore, I don't think people who live with, and clearly love, their dogs refer to themselves as "dog owners" with any negative connotations in mind.

DOG OWNERSHIP STATISTICS

In the UK, around 25 per cent of households own dogs,[2] although this varies by region, with more northern and more socioeconomically "deprived" areas tending to have higher rates of dog ownership (the relevance of this will become apparent later). In the US and Australia, dog ownership is even more popular, at around 38 per cent[3] and 40 per cent,[4] respectively. That said, in the absence of animal registers or licensing, it's pretty hard to know for sure how many of us own pets, or how many animals we have. We can only estimate from market research surveys typically funded by pet food manufacturers. Researchers often cry, "If only they would add pets to the household census surveys!" It's challenging to study the effects of a potential health intervention if you don't even know how many of them there are or where they might be. For this reason, some of the research studies now coming from places like Sweden, where virtually all dogs are registered to a household and an owner, are incredibly exciting, as we shall see later in Chapter 4.

THE THEORIES BEHIND WHY WE MIGHT EXPECT PETS TO IMPROVE OUR HEALTH

There are a number of different hypotheses as to how pets may positively affect our well-being. Firstly, there is the biophilia hypothesis, popularized in the book *Biophilia* (1984) by Edward O. Wilson, which suggests that humans possess an innate tendency to seek connections with nature and other forms of life. As animals living in our homes, our pets may fill this role directly, but as any dog walker will tell you, they can also get us out and connecting with nature.

A second hypothesized mechanism would be the physical activity hypothesis: that dogs may encourage us to get out and exercise.[5] Of course, being active is good not only for our physical health but also our mental health. There is pretty good evidence to support this hypothesis, which we will explore in Chapter 4.

The third main hypothesis put forward is that of social support[6] – that the companionship that pets provide is what may improve our health – because we know how integral having social support and friendships with other people is for mental well-being. Pets may also deliver a double whammy here, as they can also encourage us to meet more people, as we will explore in Chapter 5.

I would like to propose two more theories, developed during my studies of owners and their dogs. The first I call the happiness hypothesis. Psychologist and happiness researcher Paul Dolan postulates that there are two main components of happiness: feelings of pleasure, and the feeling that you have a purpose.[7] I would suggest that these are also the main components of what our pets give us: we feel pleasure being around them, especially when we watch them goofing around; and we have a sense of purpose, because the dog needs walking and the cat's litter box needs emptying. In support of my hypothesis are the two key themes that emerged from an online survey we recently conducted. We asked dog owners how owning a dog benefits their mental health, and the responses bore out the idea that what the ancient Greeks called *hedonia* (generating positive emotions and feelings of pleasure) and *eudaimonia* (having a sense of fulfilment and purpose) were present in their lives thanks to their dogs.

My second, and related, theory I call the mindfulness hypothesis. Despite having roots in ancient practices, the mindfulness movement, including the benefits of meditation, has been gaining both public interest and scientific evidence in recent years. When I am absorbed by worries about all the things on my to-do list, including writing this book, my go-to stress relief tactics are either a cuddle on the sofa with my dogs, breathing deeply while stroking their soft, silky ears and feeling their weight on my chest, or getting out in the fresh air together for a walk. There is something about being with dogs that brings us into the present, to focus on the now – a key component of mindfulness and living in the moment – and not on past regrets or future worries. As one interview participant told me, "You know people don't remember how to enjoy themselves like that, but dogs always do remember to enjoy themselves and have fun, and play, and enjoy the fresh air."[8]

There is also a biological foundation to support a mechanism behind pets making us feel good, with the role of a hormone called oxytocin. I will not go into detail about this, as it has been covered well elsewhere,[9] but interacting with our dogs produces oxytocin, also called "the love hormone" and known for its role in human bonding, in particular in maternal caregiving. Intriguingly, not only do we get surges in oxytocin when we look at our dogs (you know, that warm, rushing, fuzzy feeling that they are the cutest, squishiest thing ever), but we think our dogs also get it when they look at us.[10] We think that a wonderful feedback loop of loving feelings occurs between our pets and us. It may not just be in our heads, but in our hormones!

WHERE IT ALL BEGAN

The phrase "The Pet Effect", now meaning the idea that getting a pet will make you healthier and happier, seems to first appear in a 1988 study about blood pressure and heart rates when students were petting a dog,[11] and is currently promoted strongly by the media and marketing of veterinary and pet food companies. The research evidence in the field of "anthrozoology" arguably began with a study published in 1980 by Professor Erika Friedmann and colleagues.[12] Professor Friedmann (who deservedly won the International Society for Anthrozoology's Distinguished Anthrozoologist award in 2019), followed up a group of 92 white patients admitted to a large university hospital, in the late 1970s, with a diagnosis of myocardial infarction or angina pectoris, essentially heart attacks. She found that after one year, 28 per cent of the patients who did not own pets had died, compared to only 6 per cent of those who owned a pet. Because some of these pets were dogs, they wondered whether physical activity might be the reason behind the findings. However, 10 of the pet owners did not have dogs, and none of them died either. They concluded that pets were contributing to a positive social effect that could not just be attributed to walking the dog, although the study was criticized for its small sample size and limited analysis – something common to many studies in the early stages of any research field.

In 1991, Professor James Serpell, another founding researcher in this field, conducted a study where he followed up with people adopting a dog or a cat, and compared them to people living in the same geographic area who did not adopt a pet.[13] Again, the study was small but intriguing. Over the time he followed the 71 new pet owners and 26 non-pet owners, self-reported minor health problems decreased at 2 months and general health scores increased at 6 months, in both cat and dog owners initially, and these effects were still present in the dog owners at 10 months. Dog owners also appeared to trump the cat owners in terms of exercise, which we might expect. The group without pets also increased their recreational walking a little bit, but there were no significant changes in their other health measures. The study was small in terms of numbers, which means a greater chance of an accidental untrue statistical "false" finding. There was also the issue that these people self-reported their health (and may have over-egged it) rather than it being measured in a more objective way by the researchers. Still, the evidence of longitudinal changes in at least the perceived health of an individual before and after getting a pet, is pretty convincing to this day.

ECONOMIC IMPACTS OF PET-KEEPING

All this excitement about the potential benefits of pets on our health and well-being has also led to postulation that they may actually be saving us money (or, depending where you live, saving health systems money). Given some evidence that pet owners make fewer doctor's visits, savings could be in the billions.[14] It could, of course, also be argued that owning our pets also costs us a lot of money. Recently discovering a stash of over 10 years of receipts for vet bills reminds me how much money I have sunk into this "hobby" that simply baffles those rare people you meet who just don't "get" pets. Injuries caused by dog bites also cost health systems money to treat, but the money saved by having pets in society is likely far greater than the health treatment costs it incurs.[15]

ARE DOG OWNERS JUST DIFFERENT?

This all sounds grand, and it's simple to see why dog owners, pet food companies, and the media get so excited about the idea that dog owners are healthier and happier. However, what if dog owners are just fundamentally different to people who don't own dogs? Unfortunately, there is a lot of evidence to suggest there are differences between people who choose to own pets and those who do not, which we must bear in mind when we interpret the findings of any research studies. We know that in the UK, for example, dog owners are on average more likely to be from a lower socioeconomic group and poorer (in terms of education level, job type, and income) than people who don't own dogs. Cat owners in the UK conversely tend to come from a higher socioeconomic group than their counterparts.[16] Interestingly, in the USA, this situation appears to be reversed, with more affluent sociodemographic households more likely to own dogs.[17] Multiple studies suggest that dog owners are different in other ways too – and are more likely to be female, white and have school-aged children. Those children that have dogs are more likely to live with siblings.[18] Certain personality types are also more drawn to dogs over cats. For example, there is evidence that dog owners are more agreeable, less neurotic and more extroverted than cat owners.[19] The issue is, these sociodemographic and personality factors are also associated with whether a person has a disease or health condition, and may partially or completely explain it.

A classic example of a "confounding factor" is the apparent association between smoking, alcohol consumption, and lung cancer – it can look like alcohol consumption is strongly related to lung cancer, but this association disappears when smoking is accounted for. This is because smoking and alcohol consumption are themselves associated, or at least used to be when smoking was allowed in bars. In an example of confounding factors in pet data, a Californian Health Survey[20] reported that:

> "Unadjusted analyses found that children in pet-owning households were significantly healthier than children in non-owning households in terms of, for example, better general health, higher activity level, and less concern from parents regarding mood, behaviour, and learning ability. However, when estimates were adjusted using the double robust approach, the effects were smaller and no longer statistically significant. The results indicate that the benefits of owning

pets observed in this study were largely explained by confounding factors."

I am acutely aware of this issue; one of the lowest days of my research career was when I excitedly saw a very low "P-value" in my statistical analysis, suggesting I had found something significant, only to realize it suggested that seven-year-old children who owned dogs were more likely to be categorized as obese than those who didn't own dogs. Despite how impartial scientists try to be, of course we study something because we are interested in it and hope to find data to support our hypotheses. Did owning a dog really make kids overweight?! I frantically triple-checked which way I had set my categories, but it was correct. As it turns out, in this large study of 14,000 UK children, dogs were more likely to be owned by mothers scoring low in maternal education level. Separately, the children of mothers with a lower level of education were also of higher average weight. Once maternal education level had been adjusted for, there was no more association between dog ownership and children's weight as first erroneously appeared.[21] Therefore, please don't be offended if researchers ever ask you questions about your household income or educational level. You may think that these questions are not relevant to the topic of study, but without them, we cannot use the data effectively. We aren't really interested in knowing how much you, in particular, earn, but we have to adjust for these in our analyses in order to make correct conclusions.

One recent study I was involved in really hit home in this regard, in a way that we had not really considered before. We knew that people who own a pet now are likely to have owned a pet earlier in their lives, such as during childhood. Presumably, this was due to having an experience and "culture" of pet ownership within the family. In the new study, collaborators in Sweden studied identical and non-identical (fraternal) twins and compared them – a common way of assessing how much genetic contribution there is to an outcome. In this case, dog ownership turned out to be rather strongly genetically inherited. Genetics accounted for an estimated 57 per cent of differences in dog ownership among women and 51 per cent in men.[22] This was really surprising; dog ownership is less genetically determined than eye colour or weight, about the same as the personality trait of extroversion, but more than intelligence or sexual orientation.[23]

I had dogs as a kid, and thought this was why I decided to get dogs as an adult, because being exposed to them had made me like having them around, but it turns out that we really are a "dog family". Dog ownership is highly bound up in "nature" rather than in "nurture", as we might perhaps have expected. Because of this, perhaps people who own dogs are just genetically healthier in the first place?

What all this means is that longitudinal data that follows the same people over time, before and after they get a pet, is really required to work out if the pets are actually having an effect on owner health. Otherwise, there is the issue of causality to deal with – for example, do people who are more active choose to get dogs or do dogs make people more active? Does owning a dog make you more depressed or are depressed people more likely to seek out dog ownership? Even with longitudinal data, other variables can still muddle the interpretation. In the seminal study by James Serpell, the participants organically chose to adopt these animals and may have been different from the non-pet owners for other reasons that can explain their health differences, such as life stressors associated with whether they felt able to take on a pet at that time. Confounding factors can only begin to be accounted for once the datasets get quite large, which the earlier studies on this topic weren't. The challenge is that large, longitudinal studies are expensive and take a long time to conduct. In the meantime, the findings of each research study need to be carefully interpreted in light of what they can really tell us given their methodological limitations, and the few longitudinal studies that are available deserve greater faith in their findings.

When examining the evidence as a whole, the greatest confirmation of a positive effect on aspects of human health and well-being appears when looking at dog ownership. There are biologically plausible reasons for this. In other words, they make logical sense when you think about the mechanisms that may be behind any health benefits from having animals around. Dogs are (usually) social and like company, they are interactive with us, and they like to go for walks. However, rather than take the idea of the power of pets at face value and then wonder why our new puppy isn't miraculously solving all our problems, it is really important to look beyond the media hype and question the evidence – in particular, what conditions need to be met for these proposed benefits to occur. For whom, in what circumstances and in what ways might dogs benefit our health? This knowledge will strengthen our ability to harness the potential power of dogs. I will present in more detail the specific evidence regarding the effect of dogs on the physical health and activity of their owners (Chapter 4), and also on mental health and social interaction (Chapter 5). But first, I want to digress and consider the dog's point of view in all this.

CHAPTER 3

DOG NEEDS MUST COME FIRST

Before we can really start to talk about how our pets can benefit us, we need to address the issue of dog welfare. Our dogs are not just tools to be used to make us feel better and lose weight. In this chapter, we will begin to discuss some of the primary issues to consider when putting dog needs first, so that our dogs are happy and healthy too.

WHAT IS GOOD DOG WELFARE?

Traditionally, the idea of good animal welfare has focused on the absence of negative indicators, such as signs that the animal is in pain. The most common representation of animal welfare, and enshrined in UK law, are the principles of the Five Freedoms.

More recently, the importance has been raised of the need for positive indicators of welfare also;[1] it is not good enough to just be free from pain or fear. Does the animal show signs of being happy? Is it living a good life? Imagine you are looking at a field full of spring lambs. What sort of behaviours would you look for if you were assessing their welfare? Which lambs would you say are demonstrating a higher quality of life? The ones not limping? Or the ones actively skipping and frolicking around?

People's views of what constitutes good canine welfare are complex, combining aspects of physical health, mental well-being and the purpose and daily life of the animal. In a survey of over 2,000 participants from 12 countries, perceived welfare status of dogs varied significantly across 17 dog contexts and roles, from extremely low (e.g. fighting dogs, stray and feral dogs) to very high (e.g. guide dogs).[2] However,

---

THE FIVE FREEDOMS

Freedom from hunger or thirst by ready access to fresh water and a diet to maintain full health and vigour

Freedom from discomfort by providing an appropriate environment, including shelter and a comfortable resting area

Freedom from pain, injury or disease by prevention or rapid diagnosis and treatment

Freedom to express normal behaviour by providing sufficient space, proper facilities and company of the animal's own kind

Freedom from fear and distress by ensuring conditions and treatment which avoid mental suffering

others would disagree with these views. For example, some argue that it is unethical for dogs to be forced to work for us, such as in assistance or service roles, or that stray or feral dogs can have good welfare due to the freedom and autonomy they experience.

Most recently, the animal welfare theory has been updated to include the recognized importance of human–animal interaction, including the impact pet owners have on their animals (see opposite).[3]. Examples of these will be discussed further in this chapter and elsewhere in this book.

HOW DO WE KNOW IF A DOG IS HEALTHY?

Clearly, for a dog to be healthy they need to have all medical needs met, such as treatment for physical injuries or diseases. You may think this is quite straightforward to provide, but signs of pain and ill health can be difficult to recognize at times. For example, the gradual slowing down and reluctance to move of a dog with arthritis, which may be excused as "just old age".[4] In order to help notice any wayward signs, even apparently healthy dogs should regularly be checked by a veterinary surgeon – at least yearly for young dogs and every six months for older ones. A well as reactively treating ill health, taking good care of your dog also includes looking after its physical well-being proactively, such as providing preventative health care and correct nutrition.

PREVENTATIVE HEALTH CARE

Many of the infectious diseases that used to kill and maim our pet dogs on a regular basis, such as parvovirus and canine

IMPACT OF HUMAN–ANIMAL INTERACTION ON ANIMAL WELFARE

Situations where human–animal interactions may have negative welfare impacts include:

» when animals have had little or no prior human contact,
» when human presence adds to already threatening circumstances,
» when human actions are directly unpleasant, threatening and/or noxious,
» when humans' prior actions are remembered as being aversive or noxious
» and when the actions of bonded humans cause unintended harms.

In contrast, situations where human–animal interactions may have positive welfare impacts include:

» when the companionable presence of humans provides company and feelings of safety,
» when humans provide preferred foods, tactile contacts and/or training reinforcements,
» when humans participate in enjoyable routine activities or in engaging variable activities,
» when the presence of familiar humans is calming in threatening circumstances
» and when humans act to end periods of deprivation, inhibition or harm.

distemper (or rabies in countries where this is present, which it is not in the UK – yet), are now comparatively rare. This is due to reasonably high levels of vaccination against these diseases, so that the proportion of dogs protected remains sufficient to hinder spread throughout the wider canine population (herd immunity). However, localized outbreaks still do occur, especially where vaccination levels are lower. Importation of dogs from overseas is also a threat to disease status, and can cause outbreaks of new diseases not typically seen or vaccinated against (or treated for, in the case of parasites) in that country.

Vaccination can be a contentious issue. Some people argue that there is no need to vaccinate dogs, because they don't and their dogs are fine. They fail to understand that their dog is being protected by the herd immunity provided by the other owners who do vaccinate their dogs. The principle of vaccination is not so much about protecting your own health, but that of others collectively, especially those who can't be vaccinated for medical reasons. Other anti-vaccine arguments are that vaccinations can cause adverse reactions. In reality, such cases are extremely rare, don't have long-lasting effects, and are very likely coincidence that the dog developed an unrelated health condition after being vaccinated. Some advocate titre testing, which involves taking blood samples to measure antibody levels to judge whether a dog needs a booster. [5] However, not all immune reactions involve antibodies, and not all vaccines promote a response that can be measured in blood, so antibody testing is not always accurate. Furthermore, just because an animal may read high one day, that does not mean its immune levels will be sufficient in a few months' time. It can also be costly to check bloods, so is not accessible to many people to do regularly. It is also worth considering that not vaccinating may invalidate pet insurance or raise premiums, even if titre testing. Boarding kennels and training classes may also refuse to take your dog. Finally, some argue that promoting vaccination is just a way for veterinary surgeries to make money; vets retort that (if their job were all about profit and not their oath to promote animal health) they could make more money from attempting to treat, rather than preventing, these awful diseases.

Personally, I follow the standard veterinary protocols and have vaccinated my dogs throughout their lives, and have never experienced a problem or known a dog who has been adversely affected. To me, the potential risks from vaccination do not outweigh the benefits. I often wonder if, like me, vaccine opponents had been through the experience of sitting in a rescue kennel watching a puppy dying a terrible, bloody death in my arms from an easily preventable disease, they might change their opinion.

The standard advice is that dogs need two or three first vaccinations (usually beginning around six to eight weeks old, when immunity passed on from the mother has waned) and then yearly boosters. However, as studies show that immunity can last longer than a year for some diseases (up to three years), protocols used generally do not vaccinate against every single disease every year (within that yearly vaccine). If you are concerned, ask your vet what protocol they use and why – including the possibility of local disease needs – and check against the recommended guidance set by small animal veterinary associations, which regularly assess the scientific evidence.[6]

Dogs also need regular treatment to prevent fleas, ticks and worms. It is safest to acquire these from your veterinary surgery as many over-the-counter remedies are ineffective, or often cause adverse reactions, sometimes even killing animals.[7] Even if you wish to use a medication not prescribed by your vet, they can guide you through how to do this effectively and assess the pros and cons of various options.

Finally, dental health is also an important aspect of preventative healthcare. Left untreated, bad teeth can cause infections that spread to other organs in the body.[8] Many pet product suppliers tout special treats and foods that supposedly help, but there is very little scientific testing to support these claims. The most effective way to prevent dental problems sounds ridiculous, but is very simple. Clean your dog's teeth, with a toothbrush (and if needed, special canine toothpaste). Simple, but not necessarily easy!

WHAT SHOULD YOU FEED YOUR DOG?

Gone are the days when dogs were fed leftover scraps from the family meals. The pet food industry is worth billions, and owners are spoilt for choice, with prices ranging from cheap to crippling. There is not room in this book to discuss in detail which food you should feed your dog. However, there are some points you should consider when making your choices from the baffling array of options:

» You get what you pay for, sort of. The cheapest dog foods probably won't have high-quality ingredients or the correct balance of them. Having said that, just because a food is expensive, it doesn't mean it is good. Expensive food can be found anywhere, including the supermarket, but high-quality food is more likely to be found at your vet practice or pet shop.

» Diets should be labelled as "complete", should be correctly nutritionally balanced, and formulated to be suitable for their age (puppy food is different to adult food).

» Avoid colourings or additives supposed to make the food look or taste better from an owner's perspective, not the dog's (e.g. kibble shaped and coloured like peas or carrots).

» Try not to be fooled by buzz words such as "natural" and "grain free", which are usually marketing ploys and pretty meaningless. (An exception to this would be if your dog has a genuine allergy or intolerance to a grain type.)

In short, when it comes to choosing food for your pet, be careful to look past any hype and marketing, and do your own research on ingredients and processing, making any decisions in regards to your particular household and dog needs. For me, at the moment, that is buying from a well-established brand, in a convenient dry form. If you wish for your dog to enjoy a variety of taste and texture (an argument made against a homogenous pelleted diet), tricks I often use are occasionally mixing in some wet complete food, or making feeding time more interesting by using puzzle feeders or special bowls that slow down their gulping. I think the dogs enjoy this, but like many dog owners, I may be projecting my human viewpoint on to my dog's needs (known as anthropomorphism).

In addition to a move towards homemade food, and even vegetarian diets, there has been a recent explosion in popularity of feeding dogs raw or "BARF" (Biologically Appropriate Raw Food) diets,[9] based on the assumption that their close relatives, wolves, eat raw meat. Domestic dogs are arguably omnivorous scavengers and have been shown to have genetic changes that allow them to process starches (including grains),[10] so the "more appropriate" argument doesn't completely hold. Anecdotally, there are apparent health benefits for your dog from eating raw meat and bones, but no scientific study has ever supported this.[11] Despite the hype, the jury is still well and truly out as to whether raw feeding is good for dogs.

What we do have more evidence on are the potential risks of raw feeding. Like any home-prepared diet, it is hard to get the nutritional balance right, so seek advice if you wish to take that road. Some raw diets now come pre-formulated, but these are still problematic. As humans, we ate raw meat before we discovered cooking by fire, but now we rarely do, as it is likely to end badly in the bathroom at 1 a.m. There is also a big difference between the bacterial load associated with eating a fresh, warm kill (as natural predators do) and meat that was killed many weeks or months ago. Simply freezing meat does not kill all the pathogens (in particular parasites), or humans wouldn't need to cook it either. If you do wish to feed raw, please follow extremely strict hygiene protocols, with a separate preparation area and utensils, and lots of hand-washing. Worryingly, those who feed their dogs raw diets overwhelmingly believe it to be safe[12] even though the data suggests it is not.[13] One friend of mine was feeding his new puppy a raw diet, as given to them by the breeder, without even realizing it contained raw meat. Although many advocates would argue they or their dog have never been sick from it, the truth is that they were never *knowingly* sick because of what they are feeding their dog – in reality, if they got sick they would probably not know what caused it. If anybody in the household has compromised immunity, such as pregnant women, young children or the elderly, this is a particular risk to consider.

THE BIGGEST DOG WELFARE
PROBLEM: OBESITY

Yesterday I was walking back to my car after dropping my son at the school gate when I passed a beagle. Its body was twice as wide as its head as it wheezed and waddled along the pavement. Clearly (to me, at least) this dog was grossly overweight, but maybe the owner did not perceive it this way. (Of course, it is possible that the dog was already on a weight management plan and losing weight.) Obesity is becoming a huge problem, but it's not just the extreme cases like the one I saw, where quality of life is clearly compromised – clearly free from hunger, but not free from pain, injury or disease, nor free to express normal behaviour (like running). Research suggests that over 50 per cent of dogs suffer from being overweight, and the problem even afflicts the supposed examples of breed perfection we see in the show ring.[14] This is bad news, as not only does extra weight reduce the quality of life of our pets, but a study of over 50,000 dogs suggests that it can also reduce their life span by over two years due to the diseases it causes.[15]

So how do you know if your dog is overweight, or even obese? Because dog breed sizes and shapes vary so widely, it isn't possible to use height and weight to calculate a "body mass index" like we do in people. Instead, a nine-point "body condition score" (BCS) system is recommended, and there are slightly different versions for various sizes of dog.[16] Views from the side and above, combined with instructions to feel the dog, give a score from 1 (too thin) to 9 (obese – at least 40 per cent above ideal weight), with 4 or 5 being ideal. In basic principles, you should look for evidence of a waist that goes in a bit like an hourglass from above, and a nice tummy tuck underneath. The beagle I saw was easily a score 9 or even beyond. Although I would hope most people would recognize that this particular dog was an extreme condition, being a little overweight has become normalized and many owners perhaps aren't even aware that their dog is overweight.

In its most simple form, excess weight occurs because of a mismatch between energy intake (i.e. calories) and energy expenditure. In reality, obesity is a complex, multifactorial problem that is individual to the pet and owner. Diet and exercise are both important factors in this, but one myth is that a few extra walks is the solution. To test this, my colleague and world-leading obesity expert Professor Alex German and I designed a study to investigate which was more effective for doggy weight loss – diet or exercise. Alex runs a world-leading pet weight management clinic at the University of Liverpool, alongside veterinary nurse Georgia Woods. The experiment was actually filmed for a BBC TV series, and followed the progress of 13 lovely dogs (BCS between 6 and 9). Each dog was randomly selected to either be in the group that was fed a calorie-controlled diet (but instructed to exercise the same as previously), or the second group, which was instructed to increase their exercise (walks and play) by at least a third (but be fed exactly the same). Only the diet group significantly lost weight even though the exercise group did significantly increase their vigorous physical activity levels (as measured by physical activity monitors I put on their collars) and also improved their body circumference measurements a bit. Diet really is instrumental to successful weight loss, but tellingly, the owners in the first group found restricting their dog's diet far harder than the others who got to enjoy taking their dogs for more walks. Once the trial was over, all dogs were supported on a complete weight-loss journey. Georgia and Alex have helped hundreds of owners with their pet's weight loss, including Ollie the pug (overleaf).

Georgia advises that "recognizing there may be a problem with a pet's weight and swiftly seeking help from the veterinary practice gives the best chance of successful return to an ideal weight. The larger the weight problems, the harder this will be, so acting quickly is vital". As eating less can be physically and emotionally challenging for the dog, experts recommend using a special calorie-controlled diet so that the dog still feels reasonably full. This also helps to ensure that the nutritional needs of the dog are being met, as the level of restriction needed to ensure successful weight loss would usually mean that intake of some essential nutrients would be less than minimum requirements if using their normal food.[17] However, as I discovered when talking with our trial participants, dealing with the emotional demands of a pet on a calorie-controlled diet is still a challenge and requires strong support from the veterinary team. Furthermore, even though obesity is considered a very treatable disease, the effects will be lifelong, and those prone to obesity or that have been obese in the past will require careful monitoring throughout the rest of their life.

It's never too early to start to prevent pet obesity, and Georgia tells me there are now growth charts available[18] (just like we use for babies) so that we can track the growth of our pets against expected figures to ensure ideal growth and the maintenance of a healthy weight for life. Arguably, if we can

OLLIE'S STORY

Georgia first met Ollie in August 2019 after he was referred to the clinic for weight management.

"Ollie was a fairly quiet and laid-back, confident dog who was clearly adored by his two owners and their family. Ollie suffered from skin allergies and breathing issues, especially in hot weather, which was of particular concern for his owners, who love to walk. Ollie's owners had tried many different diets to try and reduce his weight in the past, but without any success – and mostly he continued to gain weight.

"Upon meeting us, Ollie's owners immediately accepted all the advice we gave to them and worked hard to put the advice into practice. At his first weight check, two weeks after I'd first met them, Ollie had lost weight. This had never happened before and his owners were very pleased. Over the last year, they have tried every day to do all they can to stick to the plan we set them and to ensure Ollie has the

correct food, together with all the activity and stimulation he needs. It's not always been easy, when the temptation to give treats has been strong for some family members, but we have worked together to find solutions for these situations that have worked for everyone.

"Now nearly a year on, Ollie is almost unrecognizable. Not only has he lost a significant amount of weight (36 per cent, over 5 kilograms), his health has improved and his whole demeanour has changed."

His owners reflected on his progress:

"We did not realize the extent of how much overweight he was, putting much of it down to him being a pug. His breathing is now perfect and causes no impact on his day or activities, even during hotter weather. He loves being active and can keep up with playtime with the children, or longer off-lead walks. He runs during walks which is a first. Before, he would spend much of his day sleeping, whereas now, he loves playtime and exercise."

prevent obesity occurring at all, we will be far more successful in maintaining healthy weights for our pets. Georgia
advises: "To prevent weight gain, owners must ensure they
are firstly feeding the correct food for their pet, and the
species, breed and age will all affect this choice. Food choice
should change over time as the pet grows, and owners need
to be alert to this fact." At age nine, Roxie my pug cross is
now on a formulated senior diet for this reason, as she was
beginning to develop a few "pudding rolls" around her chest,
despite her teeny portions. To prevent overfeeding, Georgia
suggests, "Once the correct food has been chosen, carefully
weigh out each day (not using a scoop or cup, which is unreliable) and extra foods, treats, and scraps should be limited
to 10 per cent above the total intake each day, if they are to
be given at all."

As owners, we have complete control over what our dogs
eat, so in theory obesity should not be a problem. The compli-
cation comes because we often feed our dogs for reasons other
than fuelling them sufficiently for their life. Georgia suggests
that play or grooming should also be used to replace positive
interactions that used to be food-based. Rather than reaching
for the treat pot every time you want to show your dog how
much you love them, what else can you do with them that
they enjoy?

It may help to know that pet dogs cannot adjust their
own food intake appropriately like some cats can (although
I have yet to own one – I did try feeding my cat Tabitha
ad-lib once, with a constant measure in her bowl, but I had
to stop when I began to think she would literally burst if I
didn't). In my experience, most dogs would eat and eat and
eat again if they were allowed. Brie, our pug-beagle cross

(overleaf), once secretly began eating a sack of kibble from the bottom when someone decided to temporarily store it on top of a dog crate. I only realized when she waddled into the room looking like she was about to give birth. So I really laughed when a psychologist colleague approached Alex and I wanting to do a study of whether domestic dogs adjust what they eat depending on the portion size given. In humans, the bigger the portion we are given, the more we tend to eat of it. Sure enough, we sort of proved this in dogs too.[19] I say "sort of" because in order to do the experiment, one has to weigh how much food is left over at the end of the meal. As I predicted, many of the experiments were compromised because the dog troughed the entire lot, even though we had provided very large portions at the limit of what might constitute a danger to them to be eating in one go. It may also help to know that a dog's tendency to overeat can be genetic. Some unfortunate Labradors and other retriever breeds (at least) have a specific "greedy" gene which means they are always feeling desperately hungry, the poor things.[20] So the fact that your dog wants to eat and eat isn't entirely your fault and isn't your dog's fault – it's genetic. However, all this doesn't mean owners have an excuse to let their dogs eat uncontrollably, but forewarned is forearmed, and owners of these breeds in particular can use this knowledge to plan their feeding strategy.

Clearly, I believe overfeeding is a big problem, yet I am also going to be advising you to use food rewards to train your pets. How can these two things work? I can assure you I have fed many dogs lots of food rewards, and none of them got fat. When it comes to training with food, a tiny morsel (half of a little finger-nail) is all that is needed to signal "well done!"

If you use a giant biscuit, the dog will have forgotten what it was being rewarded for by the time it has finished eating anyway. A top tip is to use portions of your dog's weighed-out daily food ration for training, rather than giving extra treats as rewards. In fact, I'm not sure that we need to feed dogs in a bowl at all, apart from for our own convenience. If need be, a dog's entire daily food ration could potentially be used as a training resource.

HOW DO WE KNOW IF A DOG IS HAPPY?

For sure, taking good care of any dog includes their emotional health (as highlighted by the goal of freedom from fear and distress). I believe that no dog should ever be forced or frightened into doing things for our benefit, and reward-based training methods should be used wherever possible (more on how to do this in Chapter 7). This is an important part of training a dog ethically, but there is also another reason. For a dog to reliably perform whatever task we have trained him to do, when we need him to, we must make sure the dog actually wants to do it. And the only way to make sure of this is if the dog enjoys doing it.

An understanding of dog behaviour is fundamental to developing this mutual relationship – in order to communicate with your dog effectively, you need to understand what he is trying to say to you, so that you can respect his wishes. Clearly, a dog who is growling, snapping or biting is feeling stressed. However, there are many more subtle behaviours that are less well known but which are nevertheless important to understand so that you can monitor the welfare of your dog, particularly during training.

BRIE THE PUGGLE

A worried or confused dog will tend to use one of four "F" strategies: flight, fight, freeze or faff. Dogs tend to do a lot of "faffing" around before the other three come into play. This can include excessive sniffing around and pretending to ignore you, known as "displacement behaviour"; i.e. "I don't know what to do so I am going to do something I know how to do well, like sniffing, peeing and barking instead." Behaviours commonly used to illustrate responses to stress or threat in dogs have been conceptualized by veterinary behaviourist Kendal Shepherd in "The Canine Ladder of Aggression".[21] Signs your dog may be feeling a bit overwhelmed are: showing visible whites of the eyes (like half-moons), nose licks, excessive yawning, ears that are flattened or low, raising a front paw, turning their head away, or rolling on their back and exposing the belly. If you don't believe me, watch two strange dogs meet and sniff each other, and you will see many of these behaviours.

SIGNS THAT A DOG IS UNHAPPY, STRESSED, FEELING THREATENED, AND MAY EVEN BECOME AGGRESSIVE

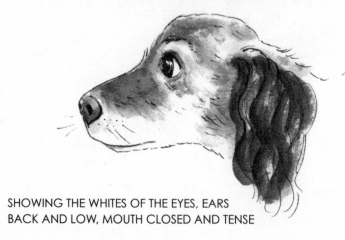

SHOWING THE WHITES OF THE EYES, EARS BACK AND LOW, MOUTH CLOSED AND TENSE

LICKING LIPS

EXCESSIVE YAWNING

RAISING FRONT PAW, EARS BACK, MOUTH STIFF, TAIL UNDER

STIFFLY LYING ON SIDE/BACK, EXPOSING BELLY

WHITES OF EYES, TURNING HEAD
AND BODY AWAY, EARS BACK

As owners, we need to learn to recognize these signals so we can head off problems before they develop. Not all dogs will show all of these behaviours, and not strictly in the same order, but they are a pretty good representation of the type of behaviours to watch out for, often termed "appeasing" or "calming" signals. Many aggression cases that I have treated that were described as "sudden" and "unpredictable" have indeed shown many of these behaviours earlier in their life or even just before the incident. Over time, the dog often learns that humans just don't listen, and may then not perform these behaviours anymore and go straight for growling or biting instead, as that is sure to get the response wanted (back off!). One example is an apparently "friendly" collie who started barking and growling at children and also ran up to and bit an adult while on a walk. When I visited the dog, it acted extremely over-friendly towards me, squirming around my feet, rolling on its back and even letting out a widdle on the floor. The owners insisted the dog was friendly to people and confident, but clearly his behaviour indicated to me otherwise. A relaxed dog who was truly not concerned by my presence would be acting a lot calmer. If dogs could talk, I want them to greet me with "Hey dude, whassup?" rather than "OH MY GOD! HELLO! HELLO! WHO ARE YOU? WHY ARE YOU HERE? DON'T HURT ME."

It can be helpful to think of these warning signals like a smoke alarm: ignore them at your peril! If you notice your dog is wide-eyed, turning their head away from something, walking away, stiffening or even growling, then take action and remove the stressor so that your dog knows these behaviours are listened to. This is the alarm system you want your dog to continue using to communicate to you that they are not happy, so that you can do something about it without the dog needing to resort to more drastic measures.

Now that we know what the signs are that a dog is clearly unhappy, what does a happy dog look like? In fact, there is no empirical research on this, because it is much easier to measure how many animals are limping than feeling good things, but I (and other people who are around dogs a lot) have a reasonably good idea what it might encompass. Don't be confused by the tail wagging, as this can also happen when the dog is in a high state of "arousal" i.e. being aggressive. Positive signs to look out for are a relaxed stance, floppy wagging tail from side to side, open mouth (not tensed), "soft" eyes, relaxed eyebrows and forehead (not furrowed), and a generally carefree attitude. The pinnacle of doggy expression of enjoyment is probably the "zoomie", when the bottom gets tucked under, head thrown back, and the dog races around in circles. Another sign of a dog attempting a positive interaction with another (be it human or dog) is the play bow (although sometimes dogs can also use this when they are confused, to suss out your reaction as friend or foe).

BEHAVIOURAL SIGNS THAT A DOG IS HAPPY OR RELAXED

FORWARD UPRIGHT EARS, RELAXED EYES, RELAXED AND OPEN MOUTH

TAIL OUT BUT NOT DIRECTLY UP, EARS FORWARD, INQUISITIVE LOOK

RELAXED, PLAYFUL, LYING ON BACK

PLAY BOW

OTHER WELFARE ISSUES – NEUTERING AND BREEDING

Finally, there are welfare risks related to breeding to consider. Neutering your animal so that it cannot reproduce has become the expected norm in many societies, and is recommended by many animal welfare organizations in the UK and USA. If you are sure you do not want your dog to be able to breed, nor do you want to have to deal with the consequences of leaving them entire (possibly roaming in males, and for females dealing with six-monthly seasons with discharge over your furniture and probably you), it is likely the right choice for you. Neutering also lowers the risk of your dog developing certain cancers, prostate problems, and uterine infections later in life. In contrast, common behavioural arguments for neutering, such as reducing aggression, are not well founded in evidence and may even make the problem worse.[22] In fact, there are some situations where neutering is not advisable, one being if you have a male dog who is very shy and fearful, and theoretically, at least, removing his confidence-boosting testosterone is the last thing he needs.

The biggest question is usually when to neuter, not if. Recommendations appear to have drifted earlier and earlier, probably due to a drive to ensure no unwanted pregnancies and gain other preventative benefits, but there is a balance to be had with ensuring sufficient hormonal development. Recent research suggests that neutering may increase the risk of joint disorders and certain other cancers, and this can vary by breed and point in time when the dog is neutered.[23] In male dogs, evidence suggests that longer exposure to gonadal hormones (i.e. neutering later) is associated with lower risk of many behavioural problems developing.[24] In females, later neutering appears to reduce the risk of some behavioural issues (including some types of fear and aggression) but increase the risks of others (chewing and howling).[25] In summary, consider the needs of your particular dog (including breed) and the current evidence when making your decision, in discussion with your veterinary surgeon.

When breeding of dogs does occur, there are more welfare risks to navigate. All breeds have inherited diseases, and although testing is available for some conditions, not all breeders use it. This is a problem because even if the breeding dogs look healthy, they could develop a disease later in life, or could be a "carrier" of the genetic mutation and pass it on unnoticed. There are also particular breeds which, in their accepted "healthy" forms, have severely compromised welfare already, in particular brachycephalic dogs (with squished faces and short muzzles). These include pugs and bulldogs, which are very popular, especially with new pet owners and young families.[26] Even though these dogs are well known to regularly snort, sneeze, snore, collapse on a walk and can't give birth naturally (they require a Caesarean section), these features are worryingly accepted as "just normal for the breed" rather than a severe deformation in the structures that allow a dog to breathe fully (see Rafa's story overleaf).[27]

At this point you may be thinking, "Hang on – doesn't Carri own pug crosses?" Yes, I do; one is a rescue dog and the other we bought as a puppy. I do love the spunky character and the friendliness of the breed – I get why they are perceived as fantastic little dogs, and my husband has a particular soft spot for short-faced dogs as well. However, the decision to own only crosses of pugs was deliberate, as I wanted my dogs to not be compromised in their ability to enjoy life and especially exercise. Both have far longer muzzles than a full pug, and even do agility training, but I am still careful not to walk them in the heat.

WHY DO WE NEED TO PUT DOGS FIRST?

From reading this chapter, you should now have a reasonable idea of what a healthy and happy dog looks like. There are two reasons why we need to put dog welfare before our own.

RAFA'S STORY

Dr Marisol Collins is a small-animal vet who is now a research colleague at the University of Liverpool. She remembers a consultation with a lovely young family who had brought their new pet for a health check. He was a friendly two-year-old French bulldog, rehomed from a relative who could no longer keep him. Marisol explained:

"Rafa was already showing the classic rasping, snorting and wheezing. Rafa had been in his new home a week and wanted to play and engage with all the attention given to him by the children, but quickly became tired and withdrawn every time. This frustrated the family, who supposed it might be for another reason, perhaps missing his previous owner, though certainly not the struggle to breathe through miniature, comma-shaped nostrils and a narrow fleshy throat, which both the relative and the internet had assured them was normal.

"We discussed how the breed-related, squashed-face features often linked to various health problems, and the ways we might need to intervene to help improve Rafa's quality of life. The family, understandably, satisfied themselves thinking Rafa just needed more time to settle at home. It wasn't until I suggested we all try an exercise – pinching our noses half shut, mouth closed, jogging on the spot for 20 seconds – which we all did (much to the amusement of colleagues looking through the consult room window), that the family, dizzy, rasping and uncomfortable, understood a

little of how playtime felt for Rafa. The family went on to have regular check-ups with Rafa, and after a time, agreed to airway surgery to improve Rafa's breathing, which it did, vastly increasing his ability to play, exercise and enjoy an active life with his family. Every time the children visited with Rafa, we had to repeat the jogging exercise in the consult room, but thankfully now without the nostril-pinching bit."

First, animals cannot help us unless their own well-being is taken care of. For example, a dog in pain cannot walk you very far, and a dog who is scared of strangers isn't going to help you make new friends in the park if it bites people who try to pet it.

The second, and more important, argument for why you should consider your dog's needs as a priority is that a dog is a living, sentient animal, not a robot. It is simply unethical (and in the UK, illegal) for us to not treat them kindly and with respect. I believe that owning a dog should be a partnership where both owner and dog benefit from each other. My interviews with dog owners have borne out that it is this mutual, reciprocal relationship, where we feel we both give to and receive love from our dogs, that engenders a sense of responsibility in owners to care for their dogs.[28] Outside observers may sometimes label the actions of dog owners as "irresponsible" if they don't agree with them. However, owner's decisions and care practices are usually based on the perception of what is "best for the dog" and with the goal of being a responsible dog owner who is doing the right thing. Thus, the good intention is there, even if sometimes the knowledge is lacking or the owner is acting on a misinformed perception.

Owning a healthy and happy dog requires a considerable amount of research, thought and intentional choice, including decisions about the physical appearance and conformation of your potential dog, before you ever lay eyes on it. This is where the information to be gleaned from this book will help. This is critical, not least because of the ethics of keeping and loving these animals, but because if our dogs aren't happy and healthy themselves, we aren't going to benefit fully from their presence in our lives.

CHAPTER 4

"THE LASSIE EFFECT"

HOW DO DOGS IMPROVE OUR PHYSICAL HEALTH?

In 2020, my dad, who is in his seventies, finally had to say goodbye to Alfie, his Jack Russell terrier (overleaf). A few weeks later, I enquired how things were going and whether he might consider getting a new dog yet. "I'm not ready. Still feel guilty about replacing him," he answered, "but the time will come. Meanwhile, there's no reason to go for a walk."

DOG OWNERS ARE MORE PHYSICALLY ACTIVE

In 2013, the American Heart Association published a statement that pet ownership, particularly dog ownership, is probably associated with decreased cardiovascular disease (CVD) risk and may have some causal role.[1] They added: "The data are most robust for a relationship between dog ownership and CVD risk reduction, particularly dog ownership and

ALFIE THE JACK RUSSELL TERRIER

increased physical activity." You will also be pleased to know that after a few months, Milo (the new rescue Jack Russell I had found them) was settling in with my dad and stepmum, providing them with a reason to walk once again through their local hills.

As we all know, though putting this into practice is a different matter, physical activity is integral to our health and well-being. It has been called the "Best Buy for Public Health" due to its effectiveness in reducing risk of cardiovascular disease, diabetes, cancers and depression. It is recommended that adults participate in a minimum of 150 minutes per week of physical activity at a moderate level of intensity.[2] However, according to government surveys, only two thirds of men and an even smaller proportion of women in the UK are actually achieving this.[3] Walking with a dog has been shown to be sufficiently intensive as a physical activity to satisfy the above criteria,[4] as long as you keep moving at a good pace. Standing in the park chatting and occasionally throwing a ball doesn't count. One hundred and fifty minutes a week is only thirty minutes per day over five days per week, which nearly all dog owners can easily achieve, surely?

Many studies now show that people who live with dogs are more likely to meet these guidelines than people without dogs. I've personally noticed that a dog walk adds around 3,000 steps to my daily count on my fitness watch. In one of my studies, I had 28 people wear accelerometers (research-grade physical activity monitors) on their waist for a week and found that the dog owners did, on average, 2,000 more steps per day (although this difference was not statistically significant, probably due to the small sample size).[5] Further-more, of the six dog owners who walked their dogs some days but not others, the difference on these days was, on average, 3,010 steps, similar to what I had noticed myself.

It is easier to get large samples of people to self-report their physical activity than wear a monitor for a week, but as you might expect, this may be subject to error due to people's guesswork or over-reporting (whether consciously or unconsciously). I got to know Dr Hayley Christian at one of the first anthrozoology conferences I attended, in Kansas City, Missouri. It took me a while to realize she was actually the Hayley Cutt, whose research on dog ownership and physical activity I had read, but she had changed her name when she married (this is a common confusion for female academics). For her PhD project in Perth, Western Australia, she had surveyed a population of 1,813 people who had moved into a new residential area, and asked them how many minutes per week they spent walking and other forms of more vigorous activity. She found that the odds of dog owners meeting the physical activity guidelines were 57 to 77 per cent higher than people without dogs.[6] At first, this excited me, and then I thought, "Hang on – if all the dog owners were walking with their dogs most days of the week, wouldn't we expect this to be higher?" As it turns out, only 23 per cent of her study participants were walking with their dog five or more times per week. After a couple of research visits to stay with Hayley in Perth (and see for myself the much more merciful weather), we designed a study to ask a UK research population the same questions.

I predicted that the difference between our dog owners and non-dog owners (the term we use for people who don't own dogs) would be greater. Firstly, in Australia the nicer weather means that there is more of a motivation for everyone to be physically active, not just the dog owners. Secondly, in the UK we don't tend to leave our dogs to exercise and chill in the garden as much (due to our different climate), but feel the need to take them out for an actual walk. At least, these were my personal hunches. Sure enough, in the UK study we found that the odds of meeting physical activity guidelines were 300 per cent higher.[7] UK dog owners walked on average seven times a week, for an average 220 minutes per week. As I discovered after a flurry of text messages and tweets pinged into my phone, the study findings were even mentioned on Friday night prime-time TV programme *The Jonathan Ross Show*. What we found in the UK is also much more than the average four times a week and 160 minutes per week of dog walking that we found in our review that summarized studies from other places such as North America, Australia and Japan.[8] In short, the data appears robust and universal, showing that adult dog owners are more physically active than people without a dog, and particularly in the UK.

However, as our cautionary tales about confounding variables in Chapter 2 highlighted, association does not imply causation; just because two variables are associated, it doesn't mean that one is causing the other. In this case, we had adjusted for many other factors and the data held. What the study design could not get round was whether dogs make people more physically active, or more active people get dogs. This is because the survey was what is known as "cross-sectional" and only measured dog ownership and physical activity at a particular point in time.

Fortunately, Hayley's study also surveyed participants at different time points. She compared 681 "continuing non-owners" with 92 new dog owners who had got a dog between the baseline measure and follow-up point one year later.[9] The difference between them was an extra 22 minutes of recreational walking in their neighbourhood per week. This was exciting and statistically significant, but actually not a lot of time. Quite tellingly, the estimated minutes kept reducing as other life changes that had occurred in the intervening period were accounted for in the analysis, suggesting again that there are big things going on in our lives that facilitate or prevent dog ownership and contribute to these supposed changes in health outcomes.

One way round this issue would be to randomly assign who is allowed to get a dog and who isn't when we study them. This is the gold standard for how clinical effects of other treatments are studied, including new drugs in a treatment group compared to a control placebo group or standard drug treatment. Of course, it is ethically impossible to do this sort of study to test the effects of dog ownership on physical activity – or is it?

This is what I thought until recently. Perhaps we could randomize an educational intervention advising people to get a dog, with a group not given this advice, but surely we couldn't actually make some people get dogs and prevent others? It seemed cruel on behalf of both the people and the dogs. That said, it has actually been tried once – in a study of hypertension treatment in high-stress professionals (stockbrokers), who were all given treatment in terms of medication. But half of them were also asked to get a cat or a dog, whose presence improved some of their hypertension measurements six months later.[10] However, that study didn't examine physical activity changes. A slightly different approach is now underway using fostering a rescue dog as a mimic for permanently acquiring a pet dog. A small pilot study group has shown increases in physical activity at six weeks, but a controlled trial (with a comparison control group, *sans* dog) has not yet been conducted.[11] The other issue with these types of studies is the matter of blinding. In drug trials, typically all the patients, clinicians and researchers are kept in the dark as to which group got the real treatment until the very end, so that data collected cannot be (deliberately or not) biased by perceptions as to what "should" be changing in whom. Clearly, it's pretty obvious, at least to the participants, if they now own a dog.

So, we are at a point where we have good reason to think that dog ownership causes increased physical activity in a general adult population, but we haven't overwhelmingly proved it, due to study design limitations. But what about other populations? Studies support dog ownership being associated with increased physical activity in older or elderly populations too.[12] However, findings for children and adolescents are more mixed. Some studies suggest that younger children may be more physically active if they own dogs.[13] This does not seem to be due simply to dog walking with their family, but also free-play unstructured time and independent mobility of the children around the local neighbourhood without their parents.[14] However, as children enter adolescence, the data supporting any positive impact of dog ownership or reported involvement in walking with the dog on overall physical activity levels is not convincing – including no differences seen in my own study of a cohort of over 2,000 young people who wore accelerometers to measure their physical activity levels repeatedly at ages 11, 13 and 15.[15]

SO DOES DOG OWNERSHIP IMPACT OWNER WEIGHT?

If dog owners are more physically active, does this play out in terms of their risk of obesity as well? Headlines may extol the virtues of dogs as the solution to the childhood obesity crisis, but as I have already noted, this does not appear to be the case,[16] nor did we find that kids with dogs scored any better on fitness tests.[17] There is a small amount of evidence to suggest that adults with dogs are less likely to be obese, but only for those dog owners who actually walk with their dog.[18] Given that our weight is a product of our calorie intake as well as expenditure, maybe it isn't surprising that a few dog walks aren't always the answer to a bulging waistline. If you want your dog to help you lose weight, you need to do a serious amount of walking.

DOG OWNERSHIP AND OTHER PHYSICAL HEALTH BENEFITS

The American Heart Association suggests that a likely benefit of physical activity with dogs is a lowering of our risk of cardiovascular disease. Professor Tove Fall in Sweden recently conducted a large study using a cleverly designed dataset. In Sweden, every member of the population has a unique ID number that is linked to their health information, such as what prescriptions they take and any diagnosed conditions. Dogs must all be registered to an owner too, and the Swedes tend to follow rules.

This means that Tove had access to a large (anonymized) dataset of 3.5 million people – she knew whether they had been diagnosed with, or died from, cardiovascular diseases, and mostly whether they lived with a dog. Due to the unique nature of this linked dataset, she also knew which breed the dogs were. Tove and her team found that living with a dog was associated with a lower risk of death from a cardiovascular cause.[19] In single-person households only (where we would postulate that it has to be you who walks the dog), dog ownership was also associated with lower risk of cardiovascular disease, and it was lowest in those who owned "hunting" dogs (terriers, retrievers and scent hounds), which we can postulate may be perceived to require lots of exercise. Supporting earlier studies such as Erika Friedmann's, they also found that dog ownership was associated with better outcomes after having suffered a major cardiovascular event.[20] Although large studies such as those conducted by Tove may be missing some of the detail we would like to know, such as participation in actual dog walking and the interactions with the pet, they allow for statistical investigations on a scale that we have never seen before, including adjustment for a wide range of confounding variables, because so much centralized data is collected, linked together and made available to researchers.[21]

And it gets even better. The data from Sweden can also be linked to pet insurance data through the unique person ID, allowing examination of what medical conditions the animals had compared to their owners, including for cats. Tove and her colleague Dr Beatrice Kennedy kindly invited me to be involved in their data analysis, which resulted in our study, published in the *British Medical Journal* 2020 Christmas Special.[22] We found that an owner was more likely to develop diabetes if their dog also had diabetes, but the same could not be said for cat ownership. This points at commonalities in health behaviours of owners and pets, possibly in terms of what they eat, but given the absence of the same association for cat owners, it seems likely to me to be driven mainly by physical activity that dogs and owners perform together.

The Swedish cohort also suggested that dog owners had a lower risk of dying within the study period from any cause, but a UK study that combined six English datasets together found no evidence of dog owners being at lower risk of death, nor death specifically from cardiovascular disease.[23] Finding studies that contradict each other is quite common in research into "The Pet Effect". A method used to try to summarize the overall evidence is to conduct what is called a "systematic" review (as opposed to a "narrative" review, in which the author can pick and choose which studies to talk about, a bit like I am doing now). If possible, all the found datasets are then combined in a super "meta-analysis" to see what the overall effect in the data is. One recently published systematic review and meta-analysis of 10 published studies did conclude that overall, dog ownership was associated with lower risk of death, mainly driven by fewer cardiovascular-related deaths.[24] Differences between study findings performed in several distinct countries may be due to varying social norms and rules about animal ownership, such as how often you must walk your dog (which Sweden stipulates must be every six hours).

PETS AND THE RISK OF ASTHMA AND ALLERGIES

Another area of physical health often talked about in relation to pets is that of asthma and allergies. Again, it has been challenging to draw firm conclusions about this topic, which is not surprising when you think about how such data would look. If family members have a history of asthma or allergies that they believe are related to pets, they are less likely to let their children own pets anyway, and therefore the data is complex to study.[25] Different types of pets may also have opposite effects. For instance, in another analysis we did on the UK birth cohort of 14,000 children mentioned previously, we found that cat ownership was associated with reduced risk of wheezing during childhood, but rabbit and rodent ownership with increased risk.[26] Even more confusingly, the effects varied depending on the type of asthma or allergies.[27] Given the evidence, researchers have concluded that it is best that "the decision of whether to keep a cat or a dog in the family should be based on arguments other than the concern of developing asthma and allergy",[28] a statement with which I would agree.

INFECTIOUS DISEASE RISKS

Zoonotic diseases (those that can pass between animals and people) are another risk to our health but, thankfully, are rather preventable. Dogs can carry and transmit many diseases, including salmonella, MRSA and, in many countries, rabies. In a study of healthy pet dogs in a community in England, we found that a quarter were shedding campylo-bacter in their faeces,[29] a bacterium that can cause human disease with food-poisoning symptoms. Luckily, most of them were a subspecies that is only thought to cause mild human symptoms, but then this suggests that perhaps many people are getting sick from their dogs without even knowing it – especially as these dogs were healthy and had no symptoms that the owners knew of. The dogs at most risk were smaller or younger, were fed commercial dog treats and human food leftovers in their bowl, and (strangely) kept alongside pet fish.

DOGS AND RISK OF PHYSICAL INJURY

When I was a baby, large enough to move and crawl around, my mum dashed upstairs for a second to fetch a fresh nappy. In the short time she was gone, I was bitten on the forehead by one of our dogs, and I still have the scar today to show for it. As was common in those days, the dog was quickly euthanized. He was part-blind, and the thought is that I may have crawled over and startled him, and thus he was deemed a risk. Mum spent the next few years concerned that I would be forever scared of dogs – I think she can rest easy on that one now. I have no memory of the incident, though I do remember once asking her where the marks on my head came from, and hearing the story.

Approximately 8,000 people per year in England suffer a dog bite serious enough to be admitted to hospital. In a recent analysis of hospital admissions data, we estimated an annual incidence in England in 2018 of 15 hospital-treated bites per 100,000 people, which has risen starkly from six per 100,000 over a 20-year period, at a rate of 4 per cent per year.[30] Even more concerning is that hospital admissions data is only the very tip of the iceberg. In my study of physical activity in dog owners versus non-dog owners, I also asked whether people had ever been bitten by a dog, and a massive 1 in 4 people had.[31] We asked for more detail about bite incidents and only a third had required any type of medical treatment, and just

0.6 per cent required hospital admission. In some ways, that is good news, but in other ways it shows how severely hospital admissions data underestimates the real numbers of dog bites occurring. Even "mild" dog bites can result in significant psychological distress for the victims,[32] let alone what dire consequences may occur for the dog.[33]

Until recently, children were the group disproportionally bitten by dogs, at least according to hospital data, but in our new analysis we have discovered that bites to adults have steadily risen and are now equal to the incidence in children. Although male victims are at higher risk of dog bites generally, we have also found that as people get older, the tables appear to turn and females become a higher-risk group. We have no idea why these differences have occurred. Perhaps changes in ownership demographics may explain it, but the answer is unlikely to be changes inherent to the dogs we own, such as their breed; why would they become more aggressive to just adults and not children, or mostly to middle-aged female adults? There are also differences in regards to the victim's age and where they get bitten; children are more likely to be bitten in the face or neck, and adults on the limbs or torso. This is likely due to physical size (children's faces are closer) but also victim behaviour. Anyone who has been round a small child will recognize research by Professor Kerstin Meints at the University of Lincoln that shows that children like to lean in and show "intrusive facial proximity" to objects, including toy animals.[34] She also suggests that young children focus on the mouth and teeth and misinterpret dogs as smiling,[35] which makes sense given that all social cues they have been taught by their family thus far are that open mouths (and seeing teeth) mean good things.

Our dogs can also be hazardous to our physical health in other ways. Although my husband insists it is the cat who is out to kill him on the stairs, our dogs can also trip us up or pull us over.[36] Elderly populations appear to be most at risk of tripping over their pets,[37] which is something to consider when wondering whether buying Granny a dog to increase her mobility is, in fact, a good idea.

Our pet dogs can also bring us into contact with other animals that injure us. In an incident known to our family as "Cowgate", my father had to be airlifted to hospital after he was attacked and trampled by cows on his dog walk near the local village. He recovered fully from the few broken ribs and bruising, but the incident prompted myself, a colleague, and an eager vet student to do some digging.

There is little scientific data on this issue. Technically, it counts as a "work-related accident", and although injuries to the public as well as farm workers should be reported to the Health and Safety Executive, that rarely happens in practice. Instead, we searched newspaper reports for evidence.[38] Bearing in mind that incidents that make the newspapers have their own biases, we found that approximately one-quarter of attacks by cows resulted in fatality and two-thirds involved dogs. I was personally aware of some guidance on this issue (probably because I move in so many dog-related circles), and when I heard what had happened I rightly suspected that my dad had not followed the official advice which is, perhaps surprisingly, to let dogs off the lead if cattle approach. Dad had understandably tried to protect little Alfie at first, and picked him up. But cows are attracted towards dogs and eventually, Dad had the sense to let go. Alfie and Sky managed to run away out of danger and after a while my dad also managed to get himself over a stone wall to safety. Fortunately, our family is now able to reminisce about this particular incident, but whilst writing this book, a family friend was tragically killed under similar circumstances. Clearly this is a serious issue, especially as letting your dog off the lead around cows goes against the typical message drilled into walkers of keeping your dog under control around livestock.

Cow phobias aside, the overwhelming evidence points to dog ownership improving our physical activity levels, and probably in turn our physical health, which may even make us live longer. The caveat is these benefits only occur if we walk our dogs, which we will address in Chapter 9 (and if we can prevent our dogs from biting us, which we will address in Chapter 6). Exercise is incredibly important for our dogs, too, providing them with both physical exertion and mental stimulation. Dog walking truly is a partnership that we both benefit from, a great example of the central theme of this book, and the key to exploiting dog ownership to improve our health and happiness.

CHAPTER 5

HOW DO DOGS IMPROVE OUR MENTAL HEALTH?

We are all well aware that good health involves both physical and mental aspects. Many of us will attest that being around dogs just makes us generally feel better and less stressed (as long as they aren't misbehaving). While writing this chapter, I put a call out on Facebook to see if any of my friends would mind me sharing their stories about how they feel their dogs have improved their mental health, and I was inundated with offers! In an actual research study, when prospective dog owners were asked what expectations of changes to their health they had for when they got their new pet, they expected increased walking (89 per cent), happiness (89 per cent) and companionship (61 per cent), and decreased stress (74 per cent) and loneliness (61 per cent).[1] As we have already discovered, they are probably right about the walking, and that exercise probably holds the key to many of the other benefits they expect, too. As if dog walking wasn't enough of a win–win situation already for the owner and the dog, it turns out to have even greater benefits. My research demonstrates that walking with dogs is a great stress reliever; when asked, people say that they are walking "for the dog", but are also quick to recount the benefits that they, the owners, get out of it as well. As one participant so concisely put it: "My friend who doesn't have her own dog comes walking with us and says that it's impossible to leave depressed after watching the dogs running around enjoying themselves."[2]

Simply watching a dog happily zooming around off-lead is a helpful way to relieve stress, but it's not the only benefit of dog walking. It can also provide us with time to reflect and work through our thoughts. I have lost count of the times people have told me that they found the answer to a problem, or the inspiration they needed to overcome a challenge, while out walking their dog. My interviews with people about their experiences of dog walking inspired a writer, Matt Black, to create a beautiful poem that makes me cry, called "Go for a Walk" (overleaf).

DO DOGS MAKE US FEEL LESS STRESSED, ANXIOUS AND DEPRESSED?

Researcher Dr Karen Allen conducted a clever study, in which she asked participants to undertake tests of their response to stressful situations, in this case a mental arithmetic task and immersing their hand in cold water. Cardiovascular reactivity (i.e. heart rate) was examined among 240 married couples, half of whom owned a pet (and half of these had dogs, and half had cats). The experiments were repeated under four randomly assigned social support treatments: alone; with pet (or friend for non-pet owners); with spouse; and with spouse and pet/friend.[3] The lowest stress responses and quickest recoveries were observed in the pet-present conditions. In other words, having the animal there made it less stressful than when on your own or with other people.

People living with chronic mental health conditions in particular report much value from owning pets, such as feeling

"GO FOR A WALK" BY MATT BLACK[4]

Go for a walk on the highest hill,
See the pink of the sun going down;
Get out and unwind, leave your worries behind,
As you stand and look over the town.

Feeling a loafer, bum stuck on the sofa?
Or it's cold? Just set off and stride.
Even half a mile will bring you a smile,
Release the endorphins inside.

Your dog and you and wide open sky,
No talk but the whisper of trees,
While the wind sorts your journey of thoughts
As answers arrive on the breeze.

More fun than a run, the gym or a swim,
A dog's joy will show you just how
Happiness habits, chasing squirrels and rabbits,
Can teach you to live in the now.

Out with a dog your spirit runs free,
The fields and the woods come alive,
You're back in the wild, like a young child
You're ruling the world, aged five.

the animals give them motivation and purpose in their lives, and non-judgemental companionship.[5] Penaran Higgs, a friend of mine and a fellow animal behaviourist, has found that animals hold an important role in both her work and personal life. When a chronic pain condition surfaced, she developed severe anxiety and felt unable to leave the house without her beloved terrier Daisydog. The dog had been by her side through the happy times, and the bad, and Pen struggled terribly when she died. But Pen rationalizes, "Immense grief is the price one pays for immense love."

Since losing Daisydog, she has rescued dogs from awful situations. She says:

"Duncandog was seriously arthritic, old and would've died in a shelter had he not come into my life. He used to start each day by sniffing my ear, and it's learning to appreciate those small things that make life wonderful! Today, I have a psychologically damaged basset who was a breeding bitch on a puppy farm. She lights up my life with her devotion to me and the kids, and her deadpan expression makes me laugh every single day! We go through this imperfect life together. I don't know how she does it, but she helps me realize that it's OK to be me."

Focus groups with dog owners without chronic mental health conditions have found similar themes. In this research, exercise with the dog was reported to increase life satisfaction; routines based around your dog can provide purpose, and doing training with your dog contributes to feelings of personal growth – all elements of the eudaimonic sense of happiness.[6] Owners also reported feelings of excitement and happiness, or calmness, around their dogs. However, they also reported negative impacts from owning dogs.

Owning a dog can be incredibly stressful at times. Professor Nancy Gee is a world-leading anthrozoologist, psychologist, and Director of the Center for Human–Animal Interaction at Virginia Commonwealth University School of Medicine. In addition to her research, she operates a therapy dog programme of 91 trained animals and their handlers, visiting the hospital system. Of all people, she knows how animals can benefit our mental health. Yet she admits to me, "I love my dogs and they bring me immeasurable amounts of joy, but right now I have to admit that the scales are leaning towards worse mental health for me than towards improved mental health." Nancy has three wonderful dogs, but each currently has different issues that are making Nancy's life challenging:

"I am massively stressed from constant worry about if/when we will lose our four-year-old poodle, Charlotte. She's currently on chemotherapy, and at one point she had lost 27 per cent of her body weight. I get so stressed over watching her eat that I actually have to walk outside at mealtime, so she doesn't sense my worry. The medical costs are astronomical, but we focus on making sure she has a good quality of life. I try to squeeze in classes for our new puppy Allie, who we got before we knew about Charlotte's medical issues, because she deserves attention, but that takes time away from Charlotte. Allie is Charlotte's full sister, so we are also concerned about Allie getting sick. Add to all this, the old guy Fletcher who suffered a head injury while playing with another dog, then developed a seizure disorder and reactivity to other dogs, which we need to deal with as well. I feel guilty about not doing enough for all three of them."

Nancy is clearly doing an outstanding job of putting her dogs' health needs first, but this run of bad luck is taking a toll on her own health.

In addition – and a key lesson to take away from this book – your dog's behaviour affects your mental well-being. Milo is a possible Saint Bernard mix who luckily found a new home where his myriad health problems were treated. His owner, Dr Karen Griffin, told me, "While I was, of course, happy that he was beginning to feel better, I could not have begun to comprehend the repercussions of a well Milo. Gone was the calm dog and in place was a youthful dog with boundless energy. I am by no stretch of the imagination a morning person, but Milo was ready to start his day in the wee hours of the morning. From the moment he awoke until late in the evening he was bursting with energy, and that quickly started to wear on me". Karen upped his exercise to try and wear him out but then other issues arose. He began "planting" and refusing to move. Ringing of the doorbell or knocking at the door was promptly followed by a flurry of panic and barking. Then there were dogs. Seeing them through the window, hearing them outside, encountering them on walks – all resulted in an intense and all-encompassing reaction. Two things became abundantly clear to Karen: "The world is a lot for Milo, and I was completely overwhelmed by him." These realizations were difficult to get her head round. She was not a novice dog owner – she had run a dog rehoming programme for years, and now she was in the middle of doing a PhD on shelter dogs. Still, she said, "I felt as though I shouldn't have been struggling to the degree that I was, but indeed it was entirely consuming me and I could not cope."

Thankfully, Karen was fortunate enough to have a wide network of people who were appropriately educated and skilled to give her the help and support that she needed. Together, they have worked hard and made significant progress, to the point where walks are enjoyable again. However, she told me:

> "Milo still has profound anxiety and remains highly reactive, and even though I am now very bonded to him, life with Milo is still extremely stressful; some days it seems unbearable and I completely break down. Life with Milo is complex and often draining. Our relationship is dynamic, as all dog-owner relationships intrinsically are. Ours just seems particularly so. We take one day at a time, continue the work that we've been doing, and take the wins as they come. Some days there are a lot of wins."

Karen's story is typical of that experienced by owners caring for a dog with unwanted behaviours. A study of the experiences of owners of a dog with a behavioural problem identified that extra time was required for management and training, it was difficult to exercise the pet, and it limited where they could go and who could visit their home.[7] It also negatively impacted their household relationships and those with family and friends. One small study concluded that those owners who saw their dogs as well behaved when left alone showed higher perceived happiness and lower perceived stress than those whose dogs misbehaved. [8] Not surprising when you think about it … a dog that is stressing you out is only going to make you feel worse, not better. As well as being a dog-welfare problem, behavioural issues are clearly not good for the owner's well-being either, and that extends to others around them. A 2019 study examined levels of the stress hormone cortisol in the hair of 58 dogs and their owners, and it suggested that the long-term stress levels in dogs and their owners are synchronized.[9] If they are stressed, we are stressed, and vice versa.

Personally, I always find the first few days of having a new dog (be it a puppy or adult rescue) a very stressful time (and I'm a professional who knows what she's doing!). I don't cope well with change and the sudden onslaught of poops, widdles, crying, barking and biting. My husband still teases me about the time I seriously suggested sending Roxie back. He has coined my reactions "post-puppy depression". The good news is that in most cases it does quickly get better, and soon you could never imagine your life without them, but if an owner is still struggling with a dog after a number of weeks, they should seek professional behavioural advice (more in Chapter 13 on where to find this help).

So, dog ownership can make us more stressed and anxious, therefore it is worth remembering that the goal of this book is not to tell you to get a dog and it will instantly improve your mental and physical health. As previously stated, one of the key concepts of this book is that the behaviour of your dog will hugely impact whether you find dog ownership mostly enjoyable. Despite many of us believing that our dogs provide us with much comfort and friendship, and impact our well-being positively, the scientific evidence to support this notion is somewhat more fuzzy. This isn't surprising when we consider that not all dogs make us feel good to be around, at least not all of the time.

We know we get surges of the feel-good hormone oxytocin around our pets (and in contrast, lowering of the stress hormone cortisol).[10] But how does this affect how we actually feel day to day? Although animals are often used in "therapeutic" interventions for special populations (such as people with disabilities – and I am not going to cover that research in this book), surprisingly few studies have examined whether dog owners are generally less anxious, stressed or depressed. Even more surprisingly, a recent study from New Zealand found that pet owners were more likely to report diagnoses of anxiety or depression, an inverse association to what we might have expected.[11] This isn't a lone finding. What studies there are tend to report dog owners being more depressed than those without a pet,[12] or no evidence of any association between dog ownership and depression,[13] including in a longitudinal study of how "positive affect" doesn't change over time in relation to getting a dog.[14]

These studies are examples of average pet owners, but what about the effect on clinical populations with a diagnosed mental health condition, where surely pets can have an impact? One promising study involved a group of "treatment-resistant" depressive patients, meaning they had received much treatment but nothing was working. It was suggested that they get a pet, and 33 out of the 80 did so. At the end of the 12 weeks, no patients in the control group had responded to their usual treatment, however, a third of the patients who got a pet had responded and their depression scores were classed as in remission.[15]

A few years ago, my friend David was diagnosed with severe anxiety and depression, and had a really rough period, which included leaving his job. After a while, he rescued Spikey, a pug he heard needed a home through a friend. David says:

"All through my different changes in medication he has been there. When I felt like I couldn't move or was zoned out, he would be next to me. The contact, when I felt distant from my emotions, was vital. Then when I started feeling a little better, he would give me routine because I knew I had to feed him and walk him at certain times of the day. Then when I was confident enough to go out into the world, the thought of him being at home waiting for me would always give me an anchor. He has given me something to live for when I felt I had nothing."

I wouldn't say that Spikey has "fixed" David's mental health struggles, but he certainly has helped him have a higher quality of life during this turbulent time. Furthermore, there are specific tasks that we can also train dogs to do (see Chapters 11 and 12), that may help people in situations like David's.

The problem, of course, with this experimental study is that the "pet/dog treatment" was not randomly assigned. There could have been other systematic differences between the people who got the pet and those who did not, which would actually account for the increased well-being – most likely that, at that moment, they were in a position in life to be able to get the animal, such as David when he was off work. As yet, there is no causal proof that simply getting a dog will make you less depressed or stressed. But what about other mental health benefits?

CAN DOGS MAKE US FEEL LESS LONELY?

One area of mental well-being where the data supporting a positive effect of pet dogs appears to be more convincing is social interactions. Arguably, the dogs themselves provide us with companionship and someone to talk to. In one study, dog owners showed greater willingness to talk to their dog about depression, jealousy, anxiety, calmness, apathy, and fear-related emotions, compared with what they were willing to share with a human confidant in the form of a friend.[16] Their willingness to disclose to their dog was similar to what they would talk about with their partner (except when talking about jealousy and apathy, which they felt more confident discussing with their dog). We all need someone to talk to about our feelings, and dogs perhaps provide that outlet for some people. Personally, I don't tend to talk to my dogs about specific things, because they are a dog and I don't expect them to understand, but I do still feel comforted by their company when I am feeling low.

Feeling lonely is a particular issue in elderly populations. In a 2014 study of 800 older adults, pet owners were 36 per cent less likely than non-pet owners to report loneliness, in a model controlling for age, living status (i.e. alone versus not alone), happy mood, and where they lived.[17] The people who lived alone and also didn't own a pet felt the most lonely of all. However, in contrast, other studies have found no difference in loneliness for pet owners compared to those without a pet.[18]

Interestingly, in Karen Allen's experimental study, people were less stressed when doing the challenging tasks in the presence of their pet than when alone, but more stressed when watched by their spouse or friend. This suggests something that I also feel I've inferred from my own research – that our relationships with animals are fundamentally different from our relationships with people. Even though we like to call them a "family member" (for want of any better term to describe it) we can't simply replace them with people (or vice versa). My pets are additional components to my life; although I feel I get emotional support from my animals, I also need my friends and family. Therefore, studying the impact of pet ownership on loneliness isn't quite straightforward, as people and animals can't be directly substituted. Furthermore, the potential impact of getting a dog on our perceived loneliness is complex to study because our dogs also affect how we interact with other people.

DOGS AS SOCIAL LUBRICANT

Owning animals makes us more sociable and more likely to interact with other people – which, again, is good for our health. Dog walkers will be well aware of the pleasant hellos we have with other dog walkers when out with our animals, but pets can also be a reason to interact with our neighbours, such as asking someone to feed your cat while you are away (or in my case, begging the people on my street to please not feed my cat, and tell them I'm very sorry he broke in through their bathroom window). It has been scientifically proven that a person is more likely to stop and talk to someone walking a dog than someone walking without a dog, even when the dog is highly trained to ensure it doesn't solicit interaction.[19] Whether the person was dressed scruffily or smartly made less of a difference than the presence of the dog. It also matters which type of dog you are walking with – a puppy or adult Labrador elicits more smiles and verbal responses than a Rottweiler.[20]

Dogs act as an icebreaker to get the conversation going. In a Canadian study of dog walkers, those that spoke to others during their dog walks reported being less lonely.[21] Many people get to know others in their community solely through dog walking, usually learning the names of the dogs before (if ever) the people! Two that spring to mind from my dog-walking past were for a long time affectionately known to us as "Labrador Man" and "R's Owner" (modified for anonymity – we did eventually learn the names of the owners).

We first met when a young Alaskan malamute decided to join us on our walk around the woods and we had to return him to his anxious owners. Despite it being very unlikely we would have ever spoken to these men if it wasn't for our shared dog-walking routines, we chatted with them every day for many years before we moved house. Conversations ranged wildly from current affairs to parenting advice for our newborn to community safety warnings. What struck me most was not only their interactions with us, but the value these men placed on their relationship with our dogs, whom they trained to race over to them (and anyone who at a distance looked remotely like them) and sit patiently for their daily treat. In fact, the hardest situation to deal with when Ben the collie was euthanized was knowing that I had to tell them. "Where's Ben?" Labrador Man innocently asked the next morning. When told the news, he responded "That's totally ruined my day", with a tear in his eye.

Therefore, it does not surprise me that pet owners have also been shown to have higher social capital, a measure of networks of relationships among people.[22] Communities with high social capital have higher civic engagement, less crime and better health. This is why former UK Prime Minister David Cameron (and subsequent politicians) are so fond of advocating that we need "Big Society" and to look after each other (some might argue, so that governments don't have to).

CAN DOGS MAKE US CLEVERER?

It is well known that when our mental health is poor and we are stressed, we also struggle cognitively to perform tasks that we would achieve more easily when our mind is in a better place. As an example, during the early anxiety of the coronavirus pandemic, while juggling home-schooling, home-working and transforming university teaching and exams to online, I managed to accidentally miss four questions out of an online survey I was building, and only realized when we went to analyze the data. This is not me; I don't make big mistakes. Thankfully, my colleagues were far more sympathetic and kind to me than I was to myself. But it raises the question: if they can make us feel less stressed, does the presence of our pets also impact our cognitive abilities?

Professor Nancy Gee, whom we met earlier, does work with preschool children that suggests that the presence of a dog when categorizing objects or remembering which objects they had seen before, can help the children restrict their attention to the demands of the cognitive tasks.[23] In a study of working memory in adults (a predictor of academic success and important to how we perform a variety of cognitive tasks), she asked college students to replicate sequences of coloured lights on a touchscreen. This was done in the presence of dogs or people, and sometimes the dog or person was not just in the room but touching the participant. The worst performance occurred when the dog was touching the person, and it was better if the dog or person was just present.[24] Perhaps animals can also be a bit distracting when we are trying to concentrate! These experiments provide some insight into the potential of animal presence to improve our thinking skills in an artificial situation. However, they do not show what effect actually owning a pet dog has on cognitive ability, which is more challenging to study, and lacking in research.

DO KIDS BENEFIT FROM GROWING UP WITH DOGS?

A lot of the research into the effect of animal presence on cognition has occurred in children, but what about other aspects of their mental health? Certainly, when I was visiting my father during the school holidays, deprived of my friends and home routine, our dog Huxley was a huge support to me. His "zoomies" running around the house and garden when I arrived made me feel incredibly loved, as did his crying outside the bathroom door and reports of moping for days when I left (of course, on reflection, maybe not such a healthy behaviour for him). Huxley was a spaniel-collie cross who stole my heart (and my bed) and was my constant companion through those long summers exploring the hillsides together, but was my mental well-being during those

HUXLEY AND ME

stressful and lonely teenage years improved by his presence? I would argue yes, but as we know, an anecdotal experience does not provide conclusive evidence.

Parents often get a pet "for the children", with the claim that it will teach them responsibility and empathy for other beings. In a collaboration with Nancy Gee, our PhD student Rebecca Purewal (Becky) conducted a systematic review of the evidence surrounding pet ownership and childhood development.[25] From 22 studies she reviewed, we could find no conclusive evidence regarding effects of pet ownership in childhood on anxiety, depression or behaviour. However, there was some evidence that pets improved self-esteem and loneliness (supporting my Huxley hypothesis). There were also some cognitive benefits, including perspective-taking abilities and intellectual development. Just like in adults, probably the greatest evidence was around increased social competence, networks and social interaction for the children who owned pets. However, study designs ranged wildly from small case studies to large cohorts, often lacked adjustment for confounding and mostly employed cross-sectional survey designs, making causation hard to prove.

Becky then analyzed the 14,000 children in the UK birth cohort I had previously examined for other outcomes, this time for emotional, behavioural and cognitive measures. She found some evidence of improved language development and pro-social behaviour in pet owners (not published yet). However, she also found that children who owned pets had poorer educational attainment, across a number of different subjects and exams, even though we had adjusted for confounding variables such as the educational attainment of the parents. As is often the case in science, our findings raised more questions than they answered.

A 2017 study using a large sample of over 5,000 Canadian children again showed strong confounding effects, which meant that initial suggestions of better general health, higher activity level, and less concern from parents regarding mood, behaviour, and learning ability, were not supported.[26] So why is the idea that pets are good for children so pervasive? It is probably due in part to selective media reporting. For example, a 2015 study was widely reported in the media. It found no evidence of an effect of dog ownership on children's body mass index, screen time, physical activity, or emotional, attention, or behavioural difficulties.[27] It did find that dog-owning children scored lower on a measure of anxiety. Guess which part of the findings was reported in news articles around the world?

WHY MIGHT DOGS NOT BE AS GOOD FOR OUR MENTAL HEALTH AS WE BELIEVE?

The anecdotal experiences extolling pet ownership as good for our mental health sound extremely convincing, but that conclusion is not fully supported by scientific evidence. There are a number of reasons why this might be the case.

The difficulty in interpreting these sorts of studies is that they are typically a cross-sectional survey of a group of people at a particular point in time, and we cannot tell whether dogs make people more depressed, or if depressed people are more likely to be drawn to dog ownership. The latter does seem perfectly plausible and could be the real explanation for many of the findings. Further complexities abound in pulling apart the meaning of these research studies, including that the effect a dog has on you might depend on who you are – one internet survey conducted in the USA found that unmarried women who live with a pet have the fewest depressive symptoms, and unmarried men who live with a pet have the most.[28]

There is also precious little data looking at the effects of dog ownership on health outcomes in the longer term, because the longer the study continues, the more it costs and the more participants lose interest in reporting back (or you can't find them). If a study were to find that dog owners were initially happier or less depressed upon getting a dog, but that effect waned over a few months, would it matter? I would like to suggest maybe not, given that it is fairly normal for other things we acquire to make us think we are happier in the short term (money, a new car, a promotion at work).[29] But then "hedonic adaptation" occurs and we get used to it and reach a more normal level of perceived well-being again. That doesn't mean that were we to lose our income, car or job (or dog) we wouldn't have a big drop in our mental well-being, at least in the short term. Many studies don't take into account how long the person has owned the dog in order to assess these sorts of fluctuations in well-being.

The "buffering hypothesis" may also help to explain why we don't see the impact of pets on well-being throughout the whole population.[30] According to this idea, the social support and the emotional benefits of pet ownership may only really come into play for individuals when they are experiencing adverse or stressful events. Population-based studies looking across large groups are not well suited to detect the influence pets may have in getting us through intermittent and individual tough times.

The main reason why dog ownership is not universally associated with mental health benefits is likely because it also really depends on the type of interactions you have with your dog. As we have discussed previously, dog ownership can be stressful, especially if your dog has behavioural problems. The best evidence for a dog's impact on our well-being – stress relief and promoting social interaction with other people – seems to come via dog walking. We also know that not everyone walks with their dog, and so they are arguably missing out.

The impact of dogs on our health may also depend on the strength or nature of the relationship that we have with them. In a study conducted by one of my veterinary science undergraduates (not published yet), we designed a survey to measure the dog owner's mental health (ratings for anxiety, depression, emotional support and companionship) and also the nature of their relationship with their dog. We found that those owners who indicated greater emotional closeness with their dog also had more anxiety and depression. In contrast, those who perceived their dog as less of a burden (scored lower on perceived costs associated with dog ownership) showed reduced anxiety and depression, and greater feelings of companionship and emotional support. If owning your dog feels burdensome and challenging, mental health factors appear to score more poorly. Again, the cross-sectional nature of the survey means that we don't know whether the people with poorer mental health sought out dog ownership and became strongly attached to their pet, those with poorer mental health found the dogs harder to look after, or being closer to your dog actually made your mental health worse.

With all of the evidence outlined in this chapter in mind, the key takeaway is that the most critical factor is not whether you own a dog, it's how you interact with your dog. The case studies I gave you, combined with the research findings, suggest that we need our dogs to be medically and behaviourally healthy if we want to gain the full benefit from owning them – in particular, we need to not feel burdened by their behaviour. It is paramount to only get a dog if you think that you are in an appropriate situation to do so – otherwise, worrying about caring for them, such as whether you can pay for the vet bills or whether they are destroying the house while you're at work, may be taxing. Once you have a dog, you also need to walk with it if you are to benefit from the stress relief that this brings. In Chapter 3, we covered the basics of your dog's health needs and how to recognize if your dog is stressed, and in future chapters we will be covering prevention (Chapters 6 and 7) and also treatment (Chapter 13) of behaviour problems. We will also discuss how to get walking with our dogs more (Chapter 9), so you can put yourself in the best position to gain those mental health benefits we all want from owning our dogs.

CHAPTER 6

PREVENTION IS BETTER THAN CURE

Is your dog going to be your beloved best friend or a source of difficulty? How do you make sure that you don't end up with the latter? I often hear the phrase "There are no bad dogs, just bad owners", but I disagree. Not that I think that problem dogs are bad dogs, nor do I think that how the owner acts doesn't contribute to an animal's behaviour; I just don't believe that being a "good" dog owner can fix all dog problems. I have worked with many owners who are essentially doing all the right things and still experiencing problems with their pet's behaviour. This misguided belief that the owner is somehow the cause of a dog's problems is an incredible source of guilt, heartache and stigma when issues do develop. I've even known of colleagues in animal rescue, while dripping with blood from a serious dog bite, being badgered by potential adopters saying, "Let me take him, he just needs a loving home." Many believe that the perfect dog can be created simply through training, but that's not quite true. Behaviour (and health) problems often have their cause way before an owner ever sets eyes on their dog. There are some dos and don'ts worth sticking to, and this chapter will run you through the most important ones.

CHOOSING YOUR DOG

First, it is important that you choose the right dog for your home and lifestyle, both in terms of breed and the dog's individual personality. Be sure to get advice on this from a variety of reliable sources, not just a friend or the internet, or even the breeder. More reliable sources may be websites of recognized organizations and charities, veterinary professionals, and books written by experts. It's easy to get it wrong, despite being well intentioned. As an example, an adolescent Siberian husky once turned up at my dog-training classes because he was proving too difficult to control. He had never been let off the lead, on the advice of the breeder, and was in the process of destroying the two-bedroom house where he lived with a young family, including a toddler. Where was I supposed to start? Serious recall and lead-walking training would be critical in order to provide him with the high levels of exercise and stimulation he needed, but realistically, the family were going to struggle to devote enough time, space and energy to a young, active dog of such a breed.

I do believe that there is a way for almost everyone to own and benefit from a dog, whatever their circumstances. There is a suitable type of dog for whatever your living situation may be, but you have to put some thought and effort into finding the right one. In the case of the husky, some initial recall training sessions with the dog on a long trailing line went well. It was lovely to see the joy on the owner's faces as they watched their dog run and play with other dogs for the first time. However, I am not sure what the long-term outcome was. As a dog trainer, I often find that no news is good news, as the nature of the job means that clients only seem to remember to get in touch when things aren't going well. Fingers crossed they had a happy ending.

It is also inadvisable to take on two siblings together, even though it may seem logical and preferable that your pup has a playmate and company. One of the early chats I had with my new neighbour Wendy involved her telling me that she was about to get two cockapoo brothers. Although excited for her (and the prospect of puppies to play with), I gently advised that this isn't thought to be a good idea. Behavioural issues often present when canine siblings are raised in the same household (typically, this means aggression towards each other, hyper-attachment to each other, or both). This is known as "littermate syndrome".[1] Of course, there is potential for friction between any two dogs living together, but because siblings are highly genetically related, they are likely to have similar personalities, including likes and dislikes, and friction can occur over food, toys or attention from their owner. Furthermore, because having two puppies at once can be challenging in terms of dedicating sufficient one-to-one time for bonding and training with each pup individually, the relationship with the owners (and obedience to their commands) can pale in significance compared to the relationship with each other.

In this case, Wendy was committed, and forewarned is at least forearmed. She worked hard to make sure that the dogs were more bonded to her than each other, and the first year went well. But sure enough, as adolescence hit, occasional serious squabbles arose between them and there were concerns for both their safety and anyone who needed to intervene. Thankfully, neutering them did seem to help tame the competitive hormones, as did making sure they had plenty of quality time apart, and all is well now. Sadly, other cases don't go so well, and experts often recommend rehoming one animal.

A second major point to make is that science tells us that genetic make-up is extremely important in terms of behaviour. The fact that particular breeds are known for their inclination to behave in a particular way (for example to herd, to point, to carry objects, to protect) gives a clue as to the strong role of genetics in dog behaviour. Genetic differences have been found to be associated with breed differences in behaviour such as trainability, attachment, attention-seeking and stranger-directed aggression.[2] This is why it is essential for prospective owners to choose the right breed for their lifestyle and needs – you can't change the genetics of your dog no matter what you do as an owner. An owner also needs to choose the right dog within that breed – to not only select the right litter but also the right individual from that litter. Some research suggests that the variation within a breed is even greater than the variation between breeds,[3] which explains why it's perfectly possible to end up with a husky that doesn't like to run (true story!) or a Labrador that isn't food motivated. You may end up upset that your new golden retriever doesn't act like your old one, even though you gave it the same name (yes, I've actually seen this too).

As a priority, you should see both parents of a puppy, in order to assess their temperament. We know that a dog's tendency for aggressive behaviour in particular is heritable, in that certain breeding lines show greater aggression.[4] I conducted a case-control study comparing case dogs who had been referred for behavioural treatment by members of the Association of Pet Behaviour Counsellors, to control dogs recruited attending the same veterinary surgeries for vaccination, whose owners would be likely to seek referral to a behaviour counsellor were they to develop a behavioural problem. [5] This meant that case and control dogs were comparably from the same population (would be referred for treatment if they had a problem, and lived in the same areas) thus any subsequent differences between them we detected were likely to be valid. Our survey responses showed that owners who did not see the mum or dad of their puppy were four times more likely to experience behavioural problems with the dog later in life, than those who had viewed both parents. Try to meet both parents if you can, or at least the mother (which is easier to do). When meeting your puppy's parents, observe them closely – are they confident, friendly and relaxed with strangers?

Also seriously consider the temperament of the individual puppy within the litter. Is it friendly but not pushy or overconfident? Although structured temperament tests of puppies tend not to perform that well in scientific testing of their validity (i.e. they don't accurately predict dog behaviour later down the line), it's sensible to perform some sort of evaluation when deciding which puppy will be yours – if you are lucky enough to have some choice in the matter. Many years ago, I, and my partner at the time, waited many months for a litter of chocolate Labrador puppies from a good breeder. We wanted a boy, and the litter had only two. We were second on the waiting list for a male. When we visited the puppies (which we did regularly), I noticed that one of the boys was highly confident: going straight up to everyone, biting ankles and running off with the end of a shoelace. The other was happy to approach and say hi, but also happy to go off and play on his own. Thankfully, the other adopters (predictably) chose the first puppy, and I got the one I wanted, who turned out to be the most fun-loving but laid-back, gentle, unfazed dog I have ever met.

Fast forward a number of years, and I was in the same situation this time with Roxie and her sister, who was much shyer. My husband fell in love with the spunky, forward pup and I put my hesitancies aside (marriage is about compromise, right?). As much as I love Roxie, her outward, spirited, barky personality has always been a challenge we have to manage, and as a pup, she was the most energetic, bitey puppy I have ever had. It felt like the only way I could get her to calm down for a moment was to stroke her belly, and nine years later, she still comes and sits on my lap for a belly rub when she wants to relax. Obviously, my sample size of two anecdotal experiences aren't conclusive evidence, but I do believe that there is some truth in the idea that although a dog's outgoing nature is a positive attribute, it would ideally be paired with a calm, assured confidence, for a pup that's easier to raise.

PUPPY OR RESCUE DOG?

"Adopt, don't shop", we are told. However, in some cases, it could be argued that the more responsible action is to source a puppy of a particular breed that you have researched and selected to best fit your lifestyle. Perhaps if you have young children and want the dog to be well socialized to them, starting with a puppy is the appropriate choice. Many of my dogs have been rescue dogs, but others I have had from a puppy. When we got Roxie, we already had two good-sized, active rescue dogs, and could only physically fit a very small third one in the boot of the car. One idea was a chihuahua, but we worried it would be too delicate to be around a clumsy collie and participate in our long daily walks. We knew that in the next few years we might have children for it to deal with, too. We settled on a puppy of a chihuahua crossed with a more robust and people-loving breed, a pug. There were good reasons for our choices, but I still felt guilty for not rescuing this time. I know full well how many dogs are sat in kennels waiting for homes. Years ago, I was the one who had to decide who lived and who died that day in order to make space for the new arrivals.

Potential adopters may also find it difficult to adopt a rescue dog in practice, even if they would prefer to and have done so in the past. I've been helping anthrozoologist and dog trainer Dr Taryn Graham look for a new rescue dog near Toronto, Canada. Similar to my observations in the UK, practically every one we can find stipulates "adult-only home" or "older children only". Taryn has a one-year-old baby, so that mostly rules someone like her out, unless her experience and credentials can help her negotiate. In fact, I actually adopted Brie the puggle when my son Brandon was only two years old. Brie (then known as Chunky-Monkey – we had to change that before it stuck) was being fostered for a rescue charity by Anne-Louise, a work colleague. Despite being very shy with strangers because of her sad past, Brie's fosterer saw great potential in her as an active family dog, and her nervousness presented as a very soft temperament with no signs of aggression. Brandon was already highly supervised around our dogs, so we knew we had safe management protocols in place. In a test visit we took to meet Brie, she avoided the scary adults but approached

Brandon happily. The rescue trusted my and Anne-Louise's professional judgement and allowed us to adopt. I am aware that I am in a very privileged situation in order to be allowed to adopt not only a dog, but a problem dog in such a context, but this is not advisable for everyone.

Don't get me wrong – having worked in rescue myself, including matching dogs with new owners, I can understand why policies are designed to try to protect both animal and human welfare, and manage concerns about liability (e.g. if a dog were to bite someone). Another reason why people seeking to adopt from animal rescue charities are often turned down is that they work, and the dog will be left alone for periods each day. Although I agree that dogs need sufficient company, I feel this policy isn't right. Both my husband and I work full time, so technically we wouldn't be allowed dogs under this policy. Yet my dogs sleep most of the day whether we are around or not, and arguably live a life of luxury, including daily long walks and weekly training classes. We also manage our schedules so that it would be extremely unusual for them to be alone for a full work day. I understand that our lifestyle still may not suit some dogs, but many would cope, especially when they have the company of other dogs. In recent years, there has also been an explosion in dog-related services, including doggy day care and dog walkers, and many people also now work flexible hours or can work from home at times.

In addition, Taryn's research has uncovered even more aspects of discrimination against potential dog owners – this time for those who do not own their own home or have a fenced-in yard.[6] This is a particular problem for incoming younger generations who cannot afford to own a home like older generations did at equivalent ages. Even if tenants can get approval from their landlord, by the time the paperwork is sorted, the potential dog has already been adopted. Taryn argues that stringent qualifications may dissuade people from adopting or force them to source dogs from less-reputable sources than breeders or long-standing rescue organizations, such as online ads, potentially leading to future behavioural issues or veterinary costs down the line.

Sure enough, we found exactly this issue in our research into why UK owners ended up adopting dogs imported from abroad. Recently, there has been a sharp rise in rescue dogs being imported from outside the adopter's country, such as stray dogs from Eastern European countries. Colleagues like Taryn, in North America and beyond, are reporting similar phenomena. Much of the veterinary community is strongly against this practice, due to the risk of importation of exotic infectious diseases, as many animals are imported without sufficient health checks. Another concern is the potential behavioural issues, such as fear and aggression, sometimes seen when street dogs are suddenly forced to conform to living within our Western homes, and many quickly attempt escape.

Yet supporters quite rightfully argue that these animals morally require our help in the same way as persons seeking asylum. In the first (and currently only) published study of over 3,000 people who adopted dogs from overseas, we found that many participants had previously been turned down from UK adoption centres because they worked or had children.[7] Others were struggling to find a dog to suit their needs, because dogs in UK centres were commonly of particular breeds or had severe behavioural problems. It is easy for me to understand how in this situation, being bombarded by social media posts of cute cross-breed dogs living in overcrowded and dirty foreign shelters, rescuing from abroad would be highly tempting. However, some adopters warned that they had been unknowingly landed with dogs with complex health or behavioural issues, with little support from the rescue organization once the money had been paid and the dog delivered. The quality of the rescue organization and the dogs they provide appears to vary widely.

I really empathize with those turned away from adopting from reputable sources. Dr Karen Griffin, whom we met earlier, published a study in 2020 evaluating the policies that rehoming organizations use.[8] She tells me:

"Organizations seem to invest considerable resources in screening potential adopters, but there is limited scientific, and sometimes logical, rationale for this. Evidence could only be found to potentially justify the inclusion of less than a quarter of the characteristics about a potential adopter that organizations report as being most important. We suggest that organizations relax their strict screening criteria. Instead we suggest they focus their resources pre-adoption on ensuring potential adopters are fully prepared for the changes in their life associated with the inclusion of a new dog in their home (i.e. ensuring adopters have appropriate expectations), and post-adoption organizations should continue to offer support as adopters adjust to having their new dog (e.g. veterinary resources, appropriate dog behavioural support). For example, rather than not allowing dogs to be rehomed with children, adopters with young children should be given necessary resources and education on how to foster and manage appropriate interactions between children and dogs, in order to mitigate the safety risk and increase the likelihood that the dog will remain in the home."

In the future, I hope to see more rescue organizations move away from blanket policies, and towards assessing the needs of an individual dog and what can be put in place for them. I also feel for those who are made to feel guilty for "shopping", not adopting. Both are valid depending on the context, as long as the breeder is reputable and responsible, as we shall now discuss further.

THE IMPORTANCE OF SOCIALIZATION

Dogs are so successful as pets because they are highly adaptable to living in different environments – if you act early enough. Research has shown that how a puppy is treated in the very early days, way before you even meet it, let alone take it home, is key in determining its behavioural development.[9] Studies performed in the 1960s (that would never pass an ethical approval committee now) deprived puppies of experiences such as contact with humans in early life, and discovered that they remain irreversibly fearful. There is a critical socialization period, thought to be between three and 12 to 14 weeks, but timing also varies by breed.[10] Even if you bring your dog home at the earliest time recommended, eight weeks, and then immediately start a vaccination protocol (which will take at least four weeks), the time window left for exposing your dog to the different sights, sounds and smells of the modern world is almost over. This exposure also needs to be gentle and rewarding, taking care not to frighten the pup, as five to 12 weeks is also a time when they have a tendency to respond to new things with fear.

Of course, if you are rescuing an older dog, you may not know its history. Otherwise, you should check that puppies are being brought up in a home environment surrounded by other intriguing species, erratic children and scary washing machines, especially before five weeks old, when the fear-response period starts. Because of this sensitive period clashing with exactly the time we upend puppies and bring them home, there may be some argument (although little scientific evidence yet) to support leaving your puppy with the breeder longer, but only if it is a good breeder doing sufficient socialization for the needs of your household environment. For example, in my case-control study of dogs referred for behaviour problems, puppies who stayed with the breeder until nine or 10 weeks old had lower odds of being a case (developing behavioural problems) than those brought home at eight weeks. Dogs from "puppy farms" (large kennels where many dogs are bred) have been shown to be more likely to have later health and behavioural problems.[11] Unfortunately, many puppies who come from these mass-producing unscrupulous breeders, are fraudulently marketed as from a lovely family home. Suspicious signs to look out for are:[12]

» selling dogs of a number of different breeds,
» mass online adverts that look similar,
» the breeder offering to drop the puppy off to your home or meet somewhere,
» visiting the puppies in a home and being told that the mother is temporarily out for a walk or at the vet,
» the seller not being interested in who you are and your situation – responsible breeders are very careful who they let have their pups.

The victims of puppy farming aren't just limited to the poor puppies and their unsuspecting owners. The breeding bitches produce litter after litter before being abandoned, or worse. These dogs, such as our puggle Brie, themselves struggle to adjust to the big scary outside world,[13] and are plagued with long-term health problems such as bad teeth from overbreeding and poor nutrition. A study comparing dogs rescued from commercial breeding establishments with dogs matched on breed, sex, age and neuter status, showed that the former breeding dogs had higher rates of health problems and fears towards dogs, people and objects.[14] Occasionally, four years on, Brie's past still haunts her – she once barked at a child until they removed their fluorescent-pink dressing gown, and she can jump if we make a sudden movement beside her when she's snoozing on the sofa.

There are useful resources available online, such as "The Puppy Contract"[15] to help you decide whether you think a breeder is legitimate and is making good choices about the future welfare of the puppy and who is able to take them home. In an ideal world, you should source puppies from a breeder who has produced dogs you know, so you can have a feel for how the puppies might turn out. Good breeders also follow detailed socialization protocols with their pups.[16] These sorts of breeders are likely to be popular and to only produce one litter per year, and so they will probably have waiting lists for months or even longer. Get your name down early! For a guide to finding a responsible breeder of a puppy in the UK, see the Puppy Pathway overleaf.

When I was at university, my dad and his family got a new puppy, Sky the border collie (overleaf). Back in those days, we were not as aware of the need for puppies to be

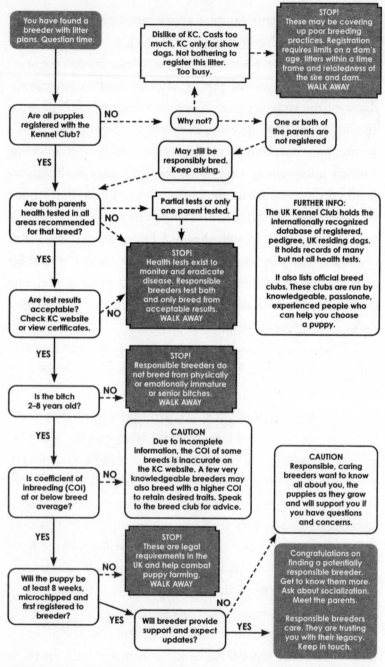

PUPPY PATHWAY GUIDE[17] A BRIEF GUIDE TO FINDING A RESPONSIBLY BRED, PEDIGREE PUPPY IN THE UK

well socialized when young, and we found out the hard way. In a turn of terrible events, the 2001 foot-and-mouth disease outbreak effectively shut down the livestock-farming areas of the country as movement restrictions tried to contain the disease from ripping through sheep and cattle populations. In remote rural areas such as where my dad lived, this meant a young Sky went from a situation of potentially at least seeing tourists and walkers occasionally strolling by the house or greeting them on her daily walks, to barely any interaction with strangers at all. Sky turned out to be not exactly friendly to strangers walking by the house – and who can blame her?

The first few weeks after you get your puppy is vital for socialization, up to and also beyond the critical period, and there are protocols and checklists available to inspire and guide you.[18] A brief guide is given overleaf.

SKY THE BORDER COLLIE, WAITING FOR HER BALL TO BE THROWN

PUPPY SOCIALIZATION NEEDS CHECKLIST

Your list of things to expose your puppy to on a regular basis[19]

NOISES

» Washing machine, tumble dryer, fan, dishwasher, microwave, etc.

» Lawnmower, vacuum cleaner, etc.

» Slamming doors

» Lorries, motorbikes, cars

» Hairdryers

» Doorbell, door knocking

» Barking dogs

» Sirens

» Fireworks

» Rain, wind, thunderstorms

» Radio, TV

» Musical instruments

PEOPLE

» Different genders

» Ages – babies, children, teenagers, adults and elderly

» All different sizes

» All different personalities

» Wearing hats, high visibility jackets, glasses, coats, costumes

» Crowds

» Joggers

» People sounds – talking loudly, crying, laughing, singing, shouting, playing

OBJECTS

» Dog toys (plastic, soft, squeaky, other textures), kids' toys

» Umbrellas

» Balloons

» Balls

» Trolleys

» Lorries, cars, bikes

» Walking aids

» Garage doors

» Bags

» Brooms

» Mirrors

» Flowers

ANIMALS

» Dogs – big, small, different breeds, puppies, older, males, females, etc.

» Cats and kittens

» Birds

» Horses

» Cows, sheep, etc.

» Chickens and ducks

» Rabbits

ENVIRONMENTS

» Veterinary surgery

» Parks

» Shops (where dogs are allowed)

» Pet shops

» Outside schools/playgrounds
» Beaches
» Car rides
» Stairs and lifts
» Going into tunnels
» Bridges (and foot bridges)
» Different surfaces – tiles, wood, grass, concrete, sand, gravel, puddles, mud
» Different heights, walking on raised surfaces

EXPERIENCES
» Being touched – held, touching paws, muzzle, ears, eyes, tail, feet, legs
» Being tied up
» Being brushed
» Bathtime
» Checking between paws
» Clipping nails
» Cleaning ears
» Teeth checking and cleaning
» Towel drying
» Wearing a harness and collar
» Walking
» Boats, public transport
» Muzzle training

However, it needs to be noted that simply experiencing things is not good enough – they must be pleasant experiences for the pup. It is essential that you don't accidentally frighten your dog in the name of "socialization". Interactions and play with other dogs needs to be carefully supervised, and with appropriate animals who are going to teach the puppy the right things. In a study of the early experience of Guide Dog puppies, surveys detailing their experiences with their puppy raisers were compared with behavioural outcomes once they reached 12 months old.[20] More experienced puppy raisers, and those who already owned other dogs, were associated with positive behavioural effects. However, those puppies that were reported to have been frightened by a person or a dog were more likely to be fearful of strangers and dogs, respectively.

Sadly, this sounds incredibly familiar to me. When we got Roxie, she was so tiny that it was easy to carry her around on our dog walks so that she could see people, dogs and the big, wide world even before her vaccination proto-col allowed her to be put on the ground to interact. As she got a little older, I also began to take the risk of walking her in places that I considered were highly unlikely to have had unvaccinated dogs around. I can't remember exactly how old she was, but around this time, my friend Kathy and I were walking the dogs one lunchtime on work premises at the veterinary school. I was walking in front with Ben the collie, and Kathy was behind me with Roxie. Somehow, a Dober-mann we were passing broke its lead or collar and rushed towards Ben, barking, and bounced off him. Both Ben and the Dobermann were a bit stunned by the unusual situation, not really knowing what to do now, and just stood there while the owner came over and apologized and retrieved her dog, and I thought nothing more of it. What I didn't realize at the time was that Roxie, a few metres behind, was paying great attention, and from her perspective, a big, scary dog had suddenly attacked her housemate. Ever since, Roxie has been nervous around many dogs, especially large or quick-moving ones, and this nervousness is expressed through a tendency to bark in alarm when she sees them. I then spent many years training and convincing Roxie that strange dogs are no threat, only for a boxer dog to rush out of an open front door and attack her from behind in the middle of the street, resulting in her requiring surgery. Not surprisingly, after that, Roxie decided "Mummy is a liar" – and we had to start again. We continue to work on her confidence around dogs and appropriate responses, and she mostly holds it together, unless it's a boxer.

There is also thought to be a secondary fear period around adolescence (eight to 12 months), when again dogs can be sensitive. I noticed this myself when training assistance dogs, as part way through their time with me, some would suddenly start to spook and bark at random things, such as a statue of a pig outside a butcher's shop in our local town (notorious to the trainers). At this time, it is worth being extra careful that your dog doesn't encounter a frightening experience that stays with them.

Finally, all my dogs go to training classes as soon as I get them and their vaccination protocols allow, even though I already have the skills in how to train them. The reason is socialization – training classes are perfect for meeting new people and dogs, and learning how to focus and listen to Mum around these distractions. Good puppy training classes fill up quickly, so get them booked in as soon as you have a date set to bring your puppy home. My dog-training colleagues are constantly having to deal with new enquiries from owners disappointed to hear that there is no space for a few months, but they already have the puppy. This is yet another reason why acquiring a puppy needs to be a process that is thought out well ahead of time and not done on a whim.

PREVENTING DOG BITES

There is one dog behaviour problem guaranteed to negatively affect the dog-owning experience, and that's if your dog bites someone, or even you. Aggression towards people is the most common issue for which behavioural and training advice is sought, even though it is probably not the most common behavioural issue dogs develop. Treating aggression is hard work, and preventing it from ever occurring is a far better approach. The strategies outlined in this book so far – sourcing and socializing your dog well, and learning to recognize and act when your dog is concerned by something – will go some way towards this, but there is more that can be done.

My interviews of dog-bite victims suggest that a common attitude is that dog bites are "just one of those things", and nothing can be done to prevent them.[21] This became particularly striking to me during my consulting work with delivery workers, who, day after day, house after house, have to navigate the high risk of a dog attack, and mostly get laughed at. In fact, the fatalistic belief that "accidents happen" is a well-known barrier to injury prevention strategies.[22] In contrast, dog experts view dog bites as very preventable. To support this hypothesis, my PhD student Dr Sara Owczarczak-

Garstecka coded behaviour that was visible on YouTube videos of people being bitten by dogs, and demonstrated that behavioural changes indicative of a bite often began to occur around 20 seconds beforehand, suggesting there is often a potential window for prevention immediately before the bite.[23]

One reason why people may think dog bites are preventable is a common perception found in my research that it was the victim's fault for doing something wrong (I called this theme "don't blame the dog"). Although there is some truth to this – that often, dog bites occur when people misinterpret the early signs that a dog is unhappy and things escalate – victim-blaming is another common barrier to injury prevention. In fact, many people are bitten in situations where it would have been difficult to do anything different, at least at the time. Examples from my research interviewees include a jogger being bitten by a dog that she didn't even know was there, a dog who redirected its reactiveness to a dog in the distance onto her walker's leg, or delivery workers posting mail through letterboxes. Our survey of people bitten by dogs demonstrated that many bites were seen as "accidental" and "unintentional", and this was more likely to be the view of owners bitten by their own dogs than of people bitten by a less familiar dog.[24]

It's clear that we have a lot of perceptions and excuses that prevent us from taking the potential risk of a dog bite seriously, especially when it comes to our own dogs. The most worrying perception we have uncovered is the belief that "It wouldn't happen to me". It's inconceivable for an owner to think that loveable Mr Fluffy would ever bite anyone, but nobody thinks their dog will bite – until it does.

Likewise, victims often think that they are good with dogs – "dogs like me" – and therefore they won't get bitten. There are also myths around dog bites that need debunking, one being that it is strange dogs, i.e. those unknown to you, that are more likely to bite you. In fact, it is well known that the majority of dog bites occur in a home, by dogs known to, if not owned by, the victim, as also found in our survey of dog-bite victims.

Another common myth is that some breeds are more likely to bite. In fact, there is no conclusive scientific evidence that a particular breed is more aggressive than others. In a systematic review we conducted of risk factors associated with human-directed dog aggression, we highlighted that breed is problematic to study both because of the high numbers needed to be able to make statistical comparisons, and often difficulties in correctly identifying a dog as a particular breed.[25] Furthermore, we found that even if a study suggests evidence of increased likelihood of aggression in a particular breed (which some do[26]), the studies don't agree on which breeds those are. Having said all this, there is some logic to the notion that some breeds may have a tendency to do more damage when they do bite, and reducing the intensity of damage when an injury occurs is also a central tenet of prevention strategies (for example, car seat belts don't stop your car from crashing but you don't get hurt as badly). Not surprisingly, being bitten by a Labrador retriever is probably going to cause more damage than a bite from a Shih Tzu.[27] Viewed from this perspective, perhaps there is some logic to the common practice of banning ownership of certain breeds, such as within the UK Dangerous Dogs Act, which bans four relatively large breeds. However, there are many breeds just as capable of causing a serious bite who are not on these lists, and breed-specific legislation is now argued to be both immoral and ineffective.[28] In fact, by labelling some breeds as "dangerous", we may have accidentally inferred that other breeds are "safe".

The message needs to be that any dog can bite, regardless of breed. This is also the view of a lesser-known UK law,[29] through which prosecution can be made against any dog perceived to be "dangerously out of control", i.e. it doesn't even have to bite, just frighten someone. This law also applies to both public and, with a recent update, private places, meaning it makes no difference if the bite occurs when the dog is defending your property (such as from a postal worker's fingers through the letterbox).

In light of this new viewpoint, there are sensible management practices that can be put in place around dogs. Postal workers can be better protected by simply installing an external letterbox, so that they don't need to come into contact with your dog at all. Likewise, baby gates should be used to keep children and dogs separated except when you are physically with them and able to closely observe and intervene (remember what happened when my mum dashed upstairs and back just to get a nappy). When they are together, they should be closely supervised at all times, meaning you are contributing to the interaction with them, not watching TV or doing the washing-up. It only takes your dog having a sore ear one day, or the child tripping and falling on a sleeping dog, to trigger an incident. It is possible to some degree to "teach" children how they should and shouldn't interact with dogs (no kissing, chasing, climbing in their bed, or approaching them while they are eating), however you cannot trust either to behave 100 per cent of the time, and young children (under seven) struggle with the cognitive abilities to follow rules or see a situation from the perspective of another (such as their dog).[30] Even though my son Brandon was brought up under strict instruction and control around dogs, I still once found him sitting in the large dog crate, surrounded by piles of dog toys and two confused-looking dogs. Luckily, their interactions had been carefully managed up until this point so that they typically found his presence pleasurable, and they tolerated my lapse and his naughtiness that one time.

People often make mistakes – "to err is human" – and we must fight the urge to trust it will all be OK, and make sure we put other protective measures in place. My student Sara conducted in-depth interviews and observations with people managing risk and safety around dogs in a number of contexts (not published yet). She found that because of the close relationships we build with dogs, even if we've only been around them for a short time, we find it easy to learn to trust them. However, this trust is based on their previous behaviour, or our experiences with dogs who look like them, and tends to override the warning signs that a dog may be exhibiting in a current situation. This explains why we often end up shocked that a bite occurred. One particular situation where this would apply is in a dog that is in pain or struggling to see or hear as they once did, as with dogs that are ageing. This means they may react differently to a situation than they have done in the past, and how we interact with our dogs and what measures we put in place to prevent bites may need to adjust over time to accommodate these new needs. I became particularly aware of this as Jasmyn got older while Brandon became a rowdy toddler. I began to notice that when we entered the house, she would often not even realize and carry on sleeping, oblivious to our presence. During this time, I made sure that the bed where she slept was in a room behind a baby gate, safely away from any risk of an unintentional reaction if she was startled.

Sara's research suggested that procedures and routines can really help to manage risk. Good habits to get into are always shutting your dog away before opening the front door to a visitor or delivery (and making sure dogs aren't ever loose in the front yard unsupervised), and feeding dogs in a place where they will not be disturbed by children or other animals. While we are talking about food, attempting to teach your dog that you can "take away his food" only serves to make him mistrust people near him while he is eating, and can create a food aggression problem, so please don't do that. Instead, if you really insist, teach him that more food is put into his bowl when people are near, rather than taken away.

Another injury prevention strategy that can be useful is making potential risk visible, and communicating it to others. This is the principle behind "yellow dog" schemes, in which dogs who are unsure of people or other dogs can wear a yellow vest, bandana or lead to indicate to people that this dog needs to be given space. While this does not replace the need for training, nor for avoiding situations that will overwhelm the dog and frighten it, it creates a visible hazard that people are more likely to notice, so that they are able to change their behaviour and avoid it. The same applies to preventing your dog from biting visitors to the house. Many bites to postal workers occur in the school holidays, when children open doors and let dogs out. Putting a clear sign outside to inform visitors that there is a dog on the premises does not absolve you of responsibility, but at least lets people know that a dog is potentially around, so that they can take precautions (such as standing further back or behind an object). This can be helpful on the odd occasion where you have shut the dog away, but then a child lets it out. (And yes, Brandon has been known to accidentally do this, too.)

Another strategy that Sara found people use to stay safe around dogs is careful regulation of their own behaviour. The idea of "emotional contagion" – that emotions and behaviours can trigger similar reactions in another – also applies to interactions between people and animals. Research participants described controlling their body and movements so as not to negatively affect the dog, such as by being calm, moving slowly and not leaning over or close to the dog. This phenomenon may partly explain why, in my community survey, I found that people who scored higher on the personality score of "emotional stability" (where higher scores indicate you are less anxious and more calm) were less likely to have ever been bitten by a dog.[31]

What you may be realizing from this section is that preventing dog bites is not about doing just one thing. In line with what is known about successful injury prevention in all sorts of other contexts, multiple approaches acting at many different levels in the system are required. There are many points in the chain which can contribute to a dog bite occurring, from the genetics of the puppy to socialization, training, household management procedures and victim behaviour. Traditional dog-bite prevention strategies tend to focus on the last point, but more effort needs to be put into the other points in order to keep people and dogs safe.

DO YOUR RESEARCH

The moral of this chapter is that you can never get this time back. Don't waste it! Do your research thoroughly: into the breed, the breeder or rescue charity, and the individual dog, including its inherited tendencies. Do everything you can to prevent health and behavioural problems, so that you don't have to deal with the consequences later.

CHAPTER 7

TO SETTEE, OR NOT TO SETTEE, THAT IS THE QUESTION

DOG TRAINING 101

WHAT KIND OF DOG DO YOU HAVE?

Some dogs really want to do things to please you. Other dogs are more concerned with what's in it for them. I have owned both types, and those that are a mixture. Those who tell you there is no need to use food as a reward in training have probably only ever owned the first type. They probably own working dogs – Labradors, spaniels, German shepherds or collies – breeds often high in "biddability", or wanting to accept and follow commands. Try getting a Chinese crested powder-puff to learn new complex assistance-dog tasks "just because it wants to". The reason for the success of many "designer cross-breeds" such as the cockapoo or Labradoodle, and why they often make great assistance dogs, is that they have tried to cross the biddability from one working breed into a breed that is highly intelligent and independent but less biddable. Most pet dogs are not working-type biddable dogs. Even if they are of a breed that traditionally works (like a Labrador or cocker spaniel), pet dogs are usually from "show" lines rather than working lines, and might as well be a different breed entirely. To train the average pet dog you need to stop expecting your dog to magically know what you want, and get smart with your training methods.

TWO KINDS OF TRAINERS

There also tend to be two types of dog trainers. The first type of dog trainer focuses on what the dog is doing wrong, and tries to stop it using "corrections" or "balance". This type often talks about the need to be a "leader" and "dominant" over your dog. Dominance is a term used by scientists to describe a relationship between individual animals, and which one, on average, tends to "win" observed interactions.

I will admit that I used to believe that I needed to be the pack leader over my dog. It was such a neat theory and, in all honesty, it made me feel powerful (which is quite reinforcing). We were told to eat before our dogs, go through doorways before them, and never let them on the couch (shock horror!). But then, like many other dog behaviourists and trainers (known as "crossover" trainers), I began to change my tune. There has been lots of recent debate over whether or not dominance even exists between domestic dogs, as well as among family groups of their much-cited relation, the wolf.

Recent data (and the scientists who came up with the theory) challenges the idea that wolves or dogs live in packs where simple hierarchies mean there is a clear "alpha" individual, as many popular training books and dramatic TV shows would have you believe.[1] The early studies were conducted on captive, unrelated wolves, which are easier to study. Distinct – and far more cooperative and less aggressive – behaviours were observed in wild wolves once they were studied. Furthermore, studies of dogs in feral situations often fail to observe hierarchies, too.

As my PhD supervisor Dr John Bradshaw writes in his book *In Defence of Dogs*:[2]

"My students and I worked at a sanctuary for rehomed dogs ... At any one time the sanctuary contains about twenty dogs, usually castrated males whose behaviour towards people is judged too unpredictable for them to be homed ... If any group of dogs had the opportunity to establish a wolf-type hierarchy, it would be at a sanctuary like this one, where the dogs have eight hours a day of unregulated interaction with their own kind ... Our study of the sanctuary dogs failed to uncover any evidence that dogs have an inclination to form anything like a wolf pack, especially when they are left to their own devices. This reinforces all the other scientific evidence indicating that domestication has stripped dogs most of the more detailed aspects of wolf behaviour, leaving behind only a propensity to prefer the company of kin to non-kin – a propensity shared by many animals and certainly not restricted to wolves or even canids. Nevertheless, many experts on dogs and dog training persist in alluding to the wolf as the essential point of reference for understanding pet dogs, despite the fact that the wolf they refer to is not the wild wolf, which values family loyalties above all else, but rather the captive wolf that finds itself in a constant battle with the unrelated wolves with which it is forced to live."

Many dog trainers also question the idea that dogs would ever view humans as other dogs in the first place, in order to dominate or be dominated by them. Do we really believe that dogs think we are also dogs? It always surprises me to think that so many people deem their dogs capable of complex thoughts, reasoning and emotion, yet incapable of realizing that their owner is not a dog.

I can understand why an owner might employ the term "dominance" when describing a situation in which they are not in control of their dog and where it may be trying to bite them. And, dare I say it, the simple fact that my own dogs mostly do as they are asked, instead of defying or challenging my leadership, means that I am, in layman's terms, "dominant" over them.[3] But I have found that when I am called in to treat an aggressive animal, usually my diagnosis is that these dogs are scared underneath. An element of fear or concern can be traced back to a point in their behavioural history, even if the aggression now looks extremely confident. They may be fearful of being in pain or being punished. They may be fearful of losing something important to them, and defensive about it. Some are frustrated that they can't get what they want or that things aren't going the way they expect. They aren't trying to establish a relationship where they are "dominant" over their owners. If they were, we might expect behaviour problems to be apparent in multiple contexts, rather than the typical one specific situation in which they are being aggressive.

There may also be confusion between the idea of dominance and personality. Note that dominance in the scientific sense is used to describe a relationship between two (or more) individuals, and is not an inherent personality trait of one individual. (Who is dominant changes depending on who is interacting). Sure, some dogs may be cheekier or pushier than others, whether genetically driven or through what they have learned they can get away with. These dogs tend to be pretty clever (behaviourists see a lot of intelligent dogs; the others usually don't cause people a problem), and with some training can quickly learn new replacement behaviours as fast as they developed the old, unwanted ones.

So if dog behaviour is not motivated by status and the desire to be dominant, then what does lie behind it? The second type of dog trainer uses "positive" or "reward-based" dog training, which instead refers to an ethos of focusing on rewarding the dog for good behaviour, and withholding rewards for bad behaviour.

REWARDS AND PUNISHMENTS

Dogs can make simple associations. One type is known as classical conditioning. The famous example of this is Pavlov's dogs, who learned to salivate at the ring of a bell that occurred whenever they were fed. My dog Jasmyn quickly learned that when the orange throw was out on the couch there was a good chance I would invite her up for a cuddle, and she would come stare at me with big eyes and a wiggly tail, in anticipation.

In classical conditioning, a dog simply makes the connection mentally between two things, but doesn't have to do anything (although they may display a reaction). Roxie and Brie have clearly associated the word "hello" with visitors arriving at the house, which now makes conference calls a bit of a nightmare.

Dogs can also make associations when learning through interactions with their environments (or us). Simply put, dogs try to repeat pleasant consequences and avoid unpleasant ones. In technical terms, this is known as the four quadrants of operant or instrumental conditioning. This can easily explain dog behaviour without the need to attribute complex plotting to dominate others and take over the world; dogs learn what works and what doesn't. In their social interactions with other dogs, or people, dogs learn what or who to avoid (for example, don't go near that dog when he has a tennis ball, he really likes it and can be rather mean around it). Dogs also like to do whatever they perceive is rewarding.

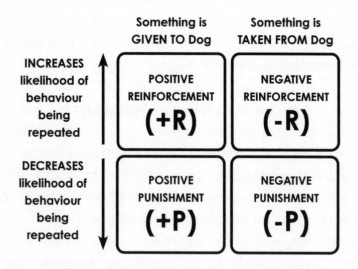

THE FOUR QUADRANTS OF OPERANT (OR INSTRUMENTAL) CONDITIONING

Rewards can be anything the dog enjoys, be it food, play, or just your attention. Positive reinforcement occurs when something is applied to the animal and as a result a behaviour increases; an example would be giving a dog a treat as a reward for sitting. If they jump up at a person and that person pets or even looks at them, the behaviour can also be rewarded. If this happens a few times, the dog will learn an association between jumping up and being reinforced for this behaviour. The next time they see a person, they will be more likely to jump up again. The environmental response (or consequence) to the action should come pretty quickly – a few seconds max – after the action, otherwise the dog will struggle to link the two. Many owners fall foul of this. The dog sits when asked, the owner fumbles around in their bag to get a treat out – and by the time the dog eats the biscuit, it is now standing up again and meanwhile has barked at a dog across the room. Which behaviour was rewarded?

Dogs will also try to avoid situations where bad things happen. For example – but not that I am advocating this – if you hit the dog in the chest when it jumped up at you, it would probably avoid doing it again. Positive punishment is the term used to describe when something is applied to the animal and as a result the behaviour decreases. As another example, if the owner shouts at them or smacks them when they have toileted in the house, next time they won't do it in front of them. They will probably still do it in the house, but will wait until the owner isn't looking. In the example I just gave, it could be argued that positive punishment was not occurring, at least for the behaviour it was intended to reduce (pooping in the house). Instead, the behaviour that was positively punished was pooping in front of the owner (not ideal). This is why using positive punishment effectively can be challenging, because it is difficult to make sure the dog understands exactly what is supposedly being punished, especially if there is a long gap between the action and the delivery of the consequence. If a dog gets accidentally rewarded once for barking, it isn't the end of the world. If it gets frightened by a punishing experience at the wrong moment, that association is going to be much harder to undo (remember Roxie's fear of dogs after one bad experience).

Negative reinforcement occurs when something uncomfortable is taken away from the dog and as a result the behaviour increases, for example, a painful, tight choke, collar that loosens when the dog isn't pulling, reinforces walking on a loose lead. Note here that when the dog is pulling, it is also being positively punished. The last quadrant is called negative punishment; a good example of this is grounding a teenager, or withholding throwing a ball for a dog who is barking. The nice thing is taken away until they can behave themselves and be positively reinforced for good behaviour. Positive reinforcement and negative punishment tend to work in tandem, and on the flip side, negative reinforcement and positive punishment tend to be used together – remember the two types of dog trainers I referred to earlier.

So in dog training terms, "positive" and "negative" actually refers to applying or taking away something, not good or bad. There is a "positive dog training" movement at the moment but this is actually a bit of a misnomer, and I prefer the term "reward-based". However, in reality it is impossible to train a dog without using all four quadrants in some form.

All four approaches may be used, but a key consideration is that there are variations in the strength of aversion used in dog training methods. For example, a harness and lead is less painful (or not at all painful) for the dog compared to a choke chain or a "prong" collar. Making a "sshht" sound, sharply yanking the dog's lead when it misbehaves, spraying water in the dog's face, or using collars that spray air in the dog's face are all attempts at positive punishment of a behaviour, despite their proponent's packaging of such methods as just "interruptions", "distractions" or "corrections". There are far less aversive ways to get your dog's attention back.

An extreme form of aversion used by some dog trainers is the electric shock collar (also called remote training collars, e-collars or electronic collars), and such devices are banned in many countries on welfare grounds. These work by delivering an electric current to the dog's neck through metal contacts, either when the trainer or owner presses a remote control button, or when the collar senses it is close to a designated area, which has a wire laid, creating a potentially invisible boundary. Although many of my contemporaries would vehemently disagree with me, I am going to put myself out there and say that personally, I am a "never say never" type, and there could arguably be situations (albeit very few) in which, on the balance of welfare benefits to the dog, having a small amount of training using a shock collar may be justified. For me, the only context that comes to my mind (though there may be others) is a situation where a strong punishment is needed to prevent the dog from ending up in much worse circumstances. One example might be if a dog is chasing livestock, and lives in a context where it would not be fair to keep the dog 100 per cent contained and kept on a lead, and the realistic alternative is being shot by a farmer. (One might argue the dog needs rehoming, but is it also fair to the owner to say they can't keep that dog, or any other?). In this context, is a brief period of training with a high-intensity, admittedly painful, aversive technique worth the benefit to the dog in having a good quality of life thereafter? It is a personal decision for sure, but could be justifiable for some.

However, research also evidences that reward-based training methods can be more effective than aversive training methods, even in this context. In a recent study by veterinary behaviourist Professor Daniel Mills at the University of Lincoln, 63 dogs with poor recall to their owner around livestock were allocated to three training groups:[4] 1) E-collar trainers (nominated by the manufacturer as the best in the field) using e-collars; 2) the same trainers following their training methods but without using e-collars; and 3) positive reinforcement trainers. They all trained the dogs for up to 150 minutes over five days, to "come" and to "sit". The final group achieved significantly more reliable and quicker responses to their commands. After doing much research into the use of e-collars, Daniel tells me:

> "The important thing about this work, is that unlike previous studies on e-collars, it did not focus on welfare. That is not because welfare isn't important, but it is easy for proponents and opponents to cherry-pick the results that support their opinion. This study focused on efficiency, and simply put, we found that e-collars were simply not necessary in this sort of context. So why would a caring owner want to use punishment when we know there are now effective alternatives? Personally, I think this study also indicates that there are different cultures among trainers,

as you mention earlier in this chapter, i.e. those that focus on correction and those that focus on shaping what you want."

No doubt there are some highly skilled trainers using electric collars out there, including the ones selected for Daniel's study. Unfortunately, most trainers who use shock collars are not highly skilled, and often use them inappropriately. For example, I was advising on some recall training for a lovely young dog who lived on a large property in a rural area. Part way through one of our sessions, the owner mentioned that the next day they were having an invisible electric fence installed around the property, so that the dog could wear a collar and wander freely without escaping. I explained my potential concerns but, ultimately, it was not up to me. If they did decide to go ahead, I advised that three things must be considered. Firstly, the dog must slowly be habituated to the collar so that he did not make any connection between wearing the collar and the shock (otherwise the dog would learn it only happened when he was wearing the collar). Secondly, care must be taken so that the dog only associated the punishment with being near the boundary area (not connecting it with being around a person, be that the owner or trainer). Most systems also have a warning buzzer sound that can be used first as a predictor of a shock if the dog ignores it. Thirdly, the collar should be initially set to a reasonably high setting, sufficiently aversive to stop the dog wanting to go near the boundary again and receive more shocks (so that the dog didn't learn to gradually tolerate higher and higher levels of pain in order to get to the rewards on the other side of the boundary). What was the outcome? The next week I returned to a distraught owner. The company who fitted the fence and collar had shoved the collar straight on the dog, and told the owner to take the dog on a lead up to the fence and repeatedly walk it through the boundary, as they increased the intensity level on the device. Not surprisingly, both the dog and the owner had a terrible experience and it was all a hugely expensive mistake. They said they were never using the collar again, and would continue to work on the recall training I was advising.

In a contrasting example, while travelling near Yosemite, my husband and I stayed in a rented RV on a piece of land near the owner's house. As is often the case in the rural United States, there were no fences around the land. I noticed that their two dogs wore electric collars and roamed freely around the property, at least up to an invisible boundary between their house and the RV, where the dogs sat patiently staring at me with big eyes, whimpers and waggly bottoms, waiting for me to go over and give them another pat. At least in fence systems, it is the animal's behaviour that controls the shock delivery, not the potential poor timing (and intermittent connection) of the handler pressing the button of a remote handheld device. The use of electric collars seems more widely accepted in the USA than the UK, although many still disagree with them. They certainly should never be recommended for owners to use for training basic skills such as recall, and often cause unnecessary pain and suffering to the dog – against their Five Freedoms.

If lots of different training methods are available and many technically work (if you are skilled enough at least), why does it matter which you choose? In the apparent words of Sigmund Freud: "Unexpressed emotions will never die. They are buried alive and will come forth later in uglier ways." I am not trying to promote the idea of psychoanalysis in non-human animals, but the point is that animal emotions are likely to get worse if not addressed properly. As I have explained already, many unwanted behaviours are caused by fear and anxiety. This means that trying to assert authority, or coerce a dog in any way, typically makes the dog more fearful and nervous in the long run (even if, temporarily, the method appears to work and suppress the behaviour). When they are trying to avoid an unpleasant outcome, dogs can quickly learn that the best defence is an offence, and this may well result in your dog resorting to aggression in the future.

In fact, studies have shown an association between the use of punishment-based and aversive training methods and the prevalence of behaviour problems, fear and weaker relation-ships.[5] One useful study, recently published by the team of Dr Anna Olsson at the University of Porto in Portugal, studied 92 dogs attending seven training schools; three that use reward-based methods and four that used aversive-based methods. Dogs from the aversive group displayed more stress-related behaviours during training, were more tense and panted more during training than dogs from the reward group.[6] They also showed higher stress-level increases (measured in their salivary cortisol) after training sessions. However, inferring causality in these types of studies is an issue. I suspect most mainly show that people with badly behaved dogs resort to punishing them, and longitudinal studies are required to provide quality evidence on the matter.

In the meantime, there is much anecdotal evidence and biological plausibility as to why it is not a good idea to train using fear and "dominance", so much so that a number of experts and charities grouped together to campaign against their use.[7] I cannot stress enough that these methods will really backfire if you are trying to train your dog to help you, and there is certainly no place for the dominance theory or aversive punishment-based methods in the type of dog training we are talking about in this book. The dog-owner relationship needs to be based on the dog knowing it is safe with the owner. You just can't build this relationship if you sometimes harm your pet.

Those who criticize my views on training methods are likely to complain that waving a piece of chicken around in front of a dog is ineffective because it is too permissive. I can reassure you that I am in no way permissive with my dogs. I do not tolerate inappropriate behaviour and I am quick to interrupt anything that I don't like so that it cannot continue to be rewarding to the dog. The most aversive I need to get is to raise my voice, and I have no qualms in doing this if my dogs step over the line. I expect my dogs to do what I ask, when I ask, and I make sure that they do. In the rest of this chapter, I will explain how to achieve this.

As we saw, research suggests that positive reinforcement training is not just kinder but is also more effective. Why? My biggest criticism of training methods that are not reward-based is that they do not show the dog what it should do. Imagine you had a little person sitting on your shoulder. In their mind, they have something they want you to do, but the only words they know that you understand are "yes" and "no". Imagine this little person shouted "no" in your ear every time you tried to do something and it wasn't what they wanted you to do. How long until you gave up and sat in a heap refusing to even try anymore? Now imagine you had a little person sitting on your other shoulder, who said "yes" every time you did the right thing. Firstly, you would feel much happier about the whole experience, and it wouldn't be long before you worked out what they wanted you to do from their clues. This is exactly what training methods feel like to our dogs. Are you a "yes" owner or a "no" owner? Which do you say to your dog the most? (Perhaps rather than "yes", you use the words "good boy/girl".) Even if we manage to stop them from doing one behaviour we don't like, they are likely to start a new one if we don't show them what they should do instead.

Rewarding the behaviours you like sounds easy in theory, but people forget to do it. In my training classes, I often observe the dogs all sitting or lying nicely and quietly, while the owners ignore them and chat to each other. Sure enough, as soon as the dogs start trying to get attention from elsewhere and bark across the room, out come the treats and the commands to sit or lie down. What is the dog learning from this situation?

In practice, a bit of both positive and negative feedback can be helpful when communicating clearly with another being, whether that be a person or dog. When interrupting your dog from doing the wrong thing, it can be helpful to have a word that means "Stop doing that right now, there is no reason to continue that behaviour as it won't be rewarded". Rather than the word "no", which has telling-off connotations, "ah-ah" can be a useful alternative to briefly get the dog's attention, so that you can now show him what he should be doing instead.

As an example, when Logan the chocolate Labrador was a puppy, I kept leaving my slippers on the floor, where he could easily get to them. Totally my fault (it's hard to change your habits) and, not surprisingly, I saw him pick them up and start to chew them. I tried to remember to put them out of reach as much as possible, but I also interrupted him from chewing and instead gave him his own toys to chew. Again and again, I swapped my slipper for a rope ragger or a dog teddy, and quietly chastised myself. After a few days, I watched Logan walk over to a slipper, look at it, look at his toy, look back to the slipper, back to the toy, and go pick up the toy and lie down and chew it. I had taught a young puppy that there was no point trying to chew my slipper, and he was supposed to chew his toy instead. Bingo!

In summary, the three golden rules of dog training are:

1. Reward the behaviour you like,

2. Ignore the behaviour you don't like,

3. Interrupt whatever behaviour you can't ignore, and show the dog what it should be doing instead.

HOW TO CHOOSE A DOG TRAINING CLASS

By now, you should hopefully have a good idea what sorts of training methods you should be looking for in a dog trainer, should you wish to work with one. If you have a young dog or puppy, I highly recommend that you attend a training class, for the socialization as well as the actual learning. The problem is, many trainers and behaviourists label their methods as reward-based, when they actually are not. Words on websites to watch out for and avoid are "natural", "leadership" or "balanced".

Even if a trainer says they use reward-based methods, this could simply mean that they sometimes give the dog a food reward, despite most of the time tugging on their necks or saying "sshht". Speak to people who have been to the classes before for their opinion on what really goes on there. Even better, go and watch a class yourself (without your dog). Do the dogs look happy to be there? Are they (and their owners) having a good time as well as learning how to behave?

Unfortunately, the dog training industry is unregulated. When choosing a dog trainer or behaviourist to work with, it is also important to check for self-proclaimed experts. "A lifetime of dog ownership" does not qualify someone as a dog trainer in the same way that having had a child does not qualify me as a midwife or a schoolteacher. Look for actual qualifications and research what these are. Are they simply a correspondence course or even just paying a fee to sign up as a member of an organization without any peer assessment? Even question claims such as "based on canine science"; if a trainer doesn't have at least a relevant master's degree, they are unlikely to have the critical thinking skills to be able to fully evaluate and understand scientific evidence, nor the means to access and keep up with the ever-changing generation of knowledge.

FINDING YOUR MOTIVATOR

Reward-based training skills are essential for training a dog to assist with your health, and creating a positive relationship. One technique you might consider is clicker training. Once the sound of the clicker device and the food reward have been "paired" using classical conditioning (this only takes a few minutes of click-treat-click-treat), the sound serves as a signal to tell your dog "it was that thing you did at the moment you heard the click that you are getting the treat for." This can be really useful in clearly communicating with your dog, especially if your dog isn't very close to you at the time you want to reward. The sound of the click can serve as a "bridge" between the task and the receiving of the reward, so that you have a few extra seconds to get and deliver a treat in the direction of the dog, without confusing it. Some trainers use a specific "marker" word instead, such as "good boy/girl" or "yes" or "clever", but this can be confusing for the dog if not taught well, as we also use such terms in daily speech around them (and then therefore have to learn not to). A clicker is a more distinct sound, but a con is you don't always have it on you to capture the behaviour you like.

Whether a clicker (or marker word) is used or not (and it isn't obligatory), for reward training to work, it is crucial to know what reward really motivates your dog. Is your dog a ball dog or a food dog, for example? My dog Jasmyn would do anything for food; I taught her to retrieve a tennis ball by cutting a hole in it and stuffing it with hot dogs. But for my collie Ben, it was all about the ball – always, any time.

Sometimes owners tell me that their dog isn't motivated by food. But I wonder if they have really tried everything. Chicken? Cheese? Hot dog? Liver? Some dogs (I'm looking at you, Labradors) would move heaven and earth for a bit of boring kibble that they eat every day. Each dog is an individual and you need to learn about what motivates yours. As an exercise, write down your dog's favourite food treats in rank order. The lowest-ranking treats are to be used for the easy tasks (such as sitting), and the power of the high-ranking treats saved for when they are truly needed (such as recall training). Without motivation, there is no power to train.

Rank	Food type
1	(e.g. cheese)
2	(e.g. hot dogs)
3	
4	
5	
6	
7	
8	
9	
10	(e.g. dried kibble)

To give you an example, Percy was a Chinese crested powder-puff I was training as a hearing assistance dog, but there was a problem. Percy couldn't give a damn if the doorbell was ringing or if the timer was bleeping. In fact, his usual response, despite my best training efforts, was to walk away from me and then cock his leg on the sofa. But then, we discovered that Percy had a food allergy, and from this point forward he could only eat one type of really boring kibble for the rest of his life. How was this going to help me motivate him? We decided that during weekdays when we trained, only his trainer – me – should ever give him this kibble. It sounds extreme, but once I had control over his only food source, which suddenly became much more important to him, the tables turned. I used most of his food for training, and only gave him food in his bowl at the end of the day if there was some left over.

I became the most amazing person in Percy's life and he would do anything I asked of him, including telling me when the doorbell was ringing.

If you have tried all sorts of potential tasty treats, and you still can't find anything your dog likes, there may be something wrong. Food is a basic need and important to dogs, and you need to check for even subtle signs of illness, pain, stress or anxiety. Also check if your dog is overweight and simply too full, or has no motivation because food is so freely available at all times (do not leave food down for them all day).

Of course, another motivator often forgotten is simply your attention. Don't forget to show your pleasure in their success! However, for learning completely new things, and especially complicated tasks, you are going to need more than just your smiles and words of encouragement. As much as we like to think they dote on us, most domestic dogs simply prefer food to petting.[8]

THE DIFFERENCE BETWEEN BRIBERY AND REWARD

Polly was a dog that I sometimes handled at the first training class I volunteered at, as I had no dog of my own at the time. She belonged to the daughter of one of the instructors, and was a Tibetan terrier, known for their rather stubborn personality. Polly was very clever and her young owner had done an amazing job of training her to perform all sorts of tasks and tricks on command – if she knew you had a hot dog for her. When training with rewards, it's rather easy to accidentally end up in this situation. There is an exercise to address this that we teach in the training classes of my colleague, behaviourist Erica Peachey, where I have taught for many years. I also use this exercise myself if I notice that one of my dogs is giving a "What's in it for me?" look when asked to do something. It teaches the dog that just because you can't see a treat, it doesn't mean it isn't coming. Note – the exercise should only be used once you are sure the dog knows the command, such as "sit" already:

1. Show the dog a treat, ask it to sit, and don't give it the treat when it sits.
2. Put all treats away, show the dog that your hands are empty, and ask it to sit. If it sits, quickly get out a treat and reward the dog.

Another strategy that can help a dog learn to work for rewards, rather than bribes, is to not reward every time. In fact, intermittent rewards are well known to be much more powerful in motivating behaviour than predictable, constant rewards – just ask a gambler. A "one-armed bandit", which sometimes gives out a reward, is far more addictive to use than a vending machine that always gives out a chocolate bar when you put the money in. Therefore, once you are sure that your dog truly understands the commands you are asking it to follow, don't give a reward every time. I take a treat bag out on my walks with my dogs, but they certainly don't get a treat every time they come back to me when called or sit when asked. Their thought pattern of "Maybe ... just maybe ... this time!" helps to keep them complying.

Another approach you can use is that of differential re-inforcement. When practising or learning a behaviour, most of the time the dog gets a small and more boring reward. But if it does a really good job, it gets a much tastier or bigger reward – it hits the jackpot! This technique can prevent your dog from thinking "What is the least I need to do to get a reward?" and therefore not trying very hard.

SETTING UP FOR SUCCESS

When training dogs, it is also important to "set up for success". The main principle to consider here is that you don't ask your dog to do anything unless you have a reasonable expectation that they will be able to do it. There is no better way for you to learn this lesson than asking your dog to sit when it is raining. Why would they want to sit and get their bottom wet?

The flip side of this approach is that if you do ask your dog to do something, you must follow through in making him do it. Otherwise, why would he bother complying in future? So now you have to stand in the rain, for as long as it takes, until, eventually, your dog sits. You asked – you now have to follow through, and next time it's raining, you won't ask them to sit for no good reason. That's a true partnership.

Setting your dog up for success also means that you need to break down tasks into easily achievable chunks, working from low to high distractions. For example, teach your dog to sit while you're both inside where it's warm, before you even think about trying to get them to sit outside in a puddle. Owners at training classes often comment that the dog can easily follow a command at home, but is being stubborn in the class and won't do it. The dog isn't being stubborn – the class environment is just far more distracting (and possibly anxiety-inducing; if you were in a room full of strangers, you might not feel comfortable lying down on the floor, either). It is important that you take note of the training environment you are using, and assess whether it is conducive to learning. If you are setting your dog up for success, you will use a relaxed and distraction-free environment (such as your living room, just you and them) when you are teaching them something for the first time, or something complicated. Only when they are confident with the response (and you are confident that they will respond when asked) would you then try asking them to do it in a slightly more distracting environment, such as in your living room with another person present, or out on a walk when there are no dogs or people around, for example. Over time, your dog will learn to generalize the command, that is, to do it in other contexts. Although to us, sit means sit, to a dog, asking them to sit inside your living room is a completely different thing to asking them to sit in an outdoor field.

EXTINCTION BURSTS

Before we finish this chapter, it is worth warning you about extinction bursts. No, I am not talking about the demise of the dinosaurs, but about what often happens when you try to ignore a well-established behaviour. Sometimes, things can get worse before they get better. Imagine a dog that has trained you to give them a chew if they bark at you while you are on the telephone (clever dog). It is worth noting here that a behaviour such as barking could be related to anxiety rather than attention seeking, but let's assume in this case that the dog isn't particularly stressed, just demanding that you pay attention to them. When you try to ignore the behaviour so that it is not rewarded, they are probably going to up the ante and bark louder and faster. This behaviour pattern is common, and known as an extinction burst.

If this happens, one strategy is to settle in for the ride, and remind yourself, "She is a dog, I am a human being. I can last longer than she can." Giving in at this stage would only reward the dog for barking even more, not less, and you need to wait until the dog stops barking before attention, or any other reward, is given. However, allowing extinction bursts to continue can also be dangerous – the dog can get very frustrated and even aggressive. This is why it is best to try to interrupt the behaviour as soon as you get an opportunity – reward the dog when it stops barking, however briefly – and redirect the dog into doing something else. Try to avoid extinction bursts from happening; think about whether there is a way that you can avoid getting in this situation again in the first place, i.e. set them up for success. The better solution to this particular issue is to give them the chew before you make the phone call (and before they bark for it), so that they are happily distracted and settled during the call. Because you have shown them what you want them to do instead, they are also beginning to learn the correct behaviour for next time the phone rings. There is also no risk of a frustrated extinction burst.

CONSISTENCY IS KEY

To settee or not to settee? That is the question. At least the one that we will conclude this chapter with. I've tried not letting dogs on the couch at all, but that was boring. I've tried having a rule where they are only allowed when invited to sit there, which worked well – especially when guests came round – until I moved in with my husband-to-be. The dogs could follow the rules, but he couldn't. Now it's a free-for-all. I've not noticed that these different approaches had any impact at all on my dogs' behaviour in general.

Conflict, such as potential aggression when you try to remove a dog from the couch, usually happens when the rules change a lot. Due to the power of intermittent rewards, if pushy or fearful behaviour around access to the comfy couch is sometimes rewarded, and at other times the dog is told off or removed, the dog will be motivated to increase the behaviour and the aggression is likely to escalate.

As we learned from sitting in the rain, consistency is key to dog training. If you ask your dog to do something, or not do something, you need to be consistent in making sure that he complies with your commands – otherwise, why would he bother? We also need to be consistent in what we are asking dogs to do. We get happy dogs and happy people when everyone knows what is expected of them and there are no surprises to argue over. If a dog knows what it needs to do, it can relax. This can be important in reducing a whole range of problems, as they are not having to second-guess their human.

Enjoy couch snuggles if you choose, but be consistent in your rules about it. The most difficult part of consistency in training is usually not the dog (who loves clear routines and expectations), but training the other humans in your house.

CHAPTER 8

COME, HEEL, WAIT

THE THREE BEST THINGS TO TRAIN YOUR DOG TO DO

Now that you have the essential training skills, we can begin to explore the full range of health-enhancing activities we can enjoy with our dogs. But what should these be, and how do we get there?

From my experiences working with dog owners over all these years, there are two main things that they really want their dogs to do, and interestingly, both relate to how nice it is to walk them. Whether their dog can sit or lie down when asked, or perform cute tricks – or not – really doesn't matter. As long as their dog comes back when they call it, so that they can enjoy letting it off the lead (Americans tend to call it a "leash"), and as long as the dog doesn't drag and pull them around when on the lead, owners tend to be mainly satisfied. Anything else is just the icing on the cake. From my experience owning dogs, there is also a third training task that makes an owner's life a lot easier if they know about it – teaching your dog to "wait". These three training tasks also encompass virtually all the commands that

Erica Peachey and I find that we use on a daily basis with our own dogs.

TRAINING YOUR DOG TO COME WHEN CALLED

In order to maximize the mental health benefits of dog walking, my research shows that dogs need to be let off the lead.[1] The walk must be enjoyable for both parties. This is where the real fun of owning a dog happens, so it is vital to find a safe place where your dog can let off steam and you can enjoy watching them. Of course, it is only possible for off-lead walking to be a relaxing exercise if your dog will come back reliably when called. For many dogs, this is one of the hardest things to train. I am going to share what I have learned over the years from training some challenging dogs, so that your dog will come when called, whatever the circumstances.

STAGE 1

Before we start on the more complicated stuff, does your dog know his or her name? This may sound like a daft question, but if your dog doesn't know when you are talking to him, how can you expect to get his attention in a distracting environment? You would be surprised how many dogs think their names are "whatsis" or "biscuits". Start by saying your dog's name and giving it a morsel of dog food, then repeat again. Then again. Soon, your dog should look at you when he hears his name, because he is expecting a tasty treat. When he does respond, praise and reward. If he doesn't respond, do not keep saying his name – that's just teaching him to ignore the sound. Instead, wave the treat right in front of his nose and lure his face towards you so that he is looking at you before you give him the treat. If your dog is not interested in the treat, find one he does like. Once again, it is your job to find out what really motivates your dog.

STAGE 2

Now that your dog knows his name and you can get his attention, we need to teach him a word that tells him to come to you. With your dog on-lead, or in a small, safe space off-lead, hold a treat down at the height of your dog's head, probably just in front of your knees (or even lower for tiny puppies). Say your dog's name, the command "come", and then walk

STAGE 2: TEACHING YOUR DOG TO FOLLOW THE TREAT TO COME TO YOU, ALSO KNOWN AS THE "PUPPY RECALL" EXERCISE

backwards. Your dog's nose should follow the treat and your dog should be following you. After a few steps, stop and give him the treat and praise. Repeat many times, walking backwards further and further, with your dog running towards you. Again, if your dog is ignoring you, stop and put the treat near his nose and lure him towards you. Don't keep repeating his name. The lead can be helpful here, but it is not to be used to pull your dog towards you. Remember, he won't be attached to you by a lead in a real-life situation. It's there to stop your dog moving any further away. Use your intelligence and motivation to get your dog to come to you. Be as exciting as possible – a high voice and silliness helps!

STAGE 3

The next training exercise requires someone to hold your dog while you walk away. It helps here if your dog is attached to a long training lead that can be left to trail, again to stop

STAGE 3: TWO-PERSON RECALL PRACTICE

him moving further away rather than for reeling him in. Get someone to hold your dog by the collar or lead while you show him the treat he can win, and then run/walk a few paces away. Then, you call your dog and the other person lets him go. Because you have left the dog, rather than the dog walking away from you, they are much more inclined to want to follow you. ("How dare my owner leave me!") Hold the treat low enough so that the dog doesn't jump at you when he arrives, and praise and reward when the dog gets to you.

STAGE 4

This exercise is similar to Stage 2, but with the addition of distractions. With your dog on a lead, approach a distraction (such as another dog) to a distance where you can still get your dog's attention. (It helps if the other dog is also on a lead, so that its owner can control it in a similar way and

STAGE 4: RECALL FROM DISTRACTION PRACTICE

it doesn't keep pestering yours during the exercise.) With a treat in your outstretched hand, call the dog away from the distraction, using your mighty owner powers and not pulling on the lead – remember, in later real life situations, the lead won't be there. Reward your dog for briefly coming away and paying attention to you instead, and then immediately let them go back to the distraction. This helps the dog learn that recall doesn't mean the end of the fun stuff, it is just a brief and worthwhile interlude.

INSTILLING THE HABIT

All these exercises have been conducted to set your dog up for practising a successful recall. In the real world, your dog will have a choice to make: either to come to you, or to run off and see the more interesting dog, squirrel or whatever other exciting distraction there might be on the other side of the park. Coming when called must be linked with a reward, and never punished, no matter how long it takes (otherwise why would they come back next time?). More importantly, it must become a habit that they do before they have really thought about it. The instinct must be for them to go to mum or dad as soon as they call. If your dog has to stop and think about whether they want to or not, you are likely to lose out. As if you are exciting enough to win a competition against a squirrel! In the UK, there is a TV advert where a child is asked which they prefer, daddy or chips ("fries" for you North Americans); the child proceeds to deliberate agonizingly over the decision. I like to imagine the same situation happening in most dog's brains as they look at you and at the distraction … daddy or chips? Mummy or that other dog? If a dog is deliberating over this choice, you have already lost the game.

The trick is for their decision to be automatic, without the need for thought. Thus, for successful recall, it is crucial to practise, practise and practise again.

SETTING UP FOR SUCCESS

The more times you call them and they do not return, the more times they are learning to ignore you – exactly the opposite of our goal. For this reason, only call your dog when you think you have at least a 90 per cent chance of them responding. This means, at the beginning of training, if they are playing with another dog, don't bother to call them away. Once your dog has learnt the meaning of the command "come", you need to proof it. "Proofing it" means you practise in gradually more difficult situations and, over time, your dog learns to be called away from ever greater distractions, until they respond even when playing with a dog. If in doubt, keep your dog on a lead or on a long trailing line at first, so that they can be managed. Remember the principle of setting a dog up for success – so don't call your dog when it's unreasonable to expect them to be able to respond, such as immediately after they have been let off the lead. Another good way to set them up for success is to call them to you when they are already on their way, not just when they don't want to come, so that they are learning the association between being called and coming to you.

MASTER LEVEL

If you threw the ball for your dog to fetch, and then called him back before he reached it, would he leave the ball and come back to you? Not many dogs would, yet this is exactly what we expect them to do when they are chasing a jogger, cat, dog, etc. It is a great deal to ask of them, especially if we have not taught them the rules of the game beforehand, in play. Some dogs have an extremely strong chase instinct. Collies and shepherds often fall into this category. It is not possible to stop dogs chasing, but it is possible to control the chase. Your dog needs to understand the rules of this game, just as with any other. The main rule is that when he is chasing something, he should want to come back to you if you call him. We want to teach him to leave what he is chasing and return when you call, so teach him this in play first, using the "chase recall" exercise (below).

CHASE RECALL

Have two toys for your dog. He should like them both, but you should know which one he prefers. Call him to you and get him to play with the less interesting toy.

1. Play with him with this toy; you throw it and he runs after it and brings it back to you (more on how to train this later, if your dog doesn't naturally do it). You then reward him and take the toy from him and continue to play.

2. After a few repetitions, throw this less interesting toy without telling him to get it, but in such a way that he will run after it. As he is running towards it, before he gets to it, call him back to you.

3. At the same time, prevent him from reaching the toy. This can be done in one of two ways. Preferably, have another person who catches and holds the toy. If this is not possible, you can have a long lead attached to his harness, which you can then grab. (If you have a large, strong or fast dog, be careful that you do not hurt yourself or your dog.)

4. As your dog turns back to look at you, throw the other (even better) toy in the opposite direction, i.e., behind you. Encourage him to chase and play with this toy.

This way, you are rewarding him for giving up on chasing one thing by letting him chase something better. You are teaching him that even when he is chasing something, if you call him, he should go back to you, as you have something even better for him. You can follow the same idea using treats if your dog is less keen on toys. You will need less space, but your reactions must be quicker.

Try this at home first, and progress to having this control when you are out on walks.

WHISTLE TRAINING

Often, if a dog has poor recall, the owner buys a dog whistle. While whistles can be helpful, they won't miraculously solve the recall problems, and they still require training to use effectively.

WHISTLE TRAINING

We talk to our dogs a lot. They hear our voices hundreds of times a day – and many times, we say their name and don't always give them a reward for responding to their name. It is not surprising that sometimes, when out on a walk and distracted by wonderful smells and other animals, they do not necessarily notice us calling them or want to respond right away, even if under more boring circumstances they can be quite good at coming when called for a reward. The blow of the whistle is a highly distinct sound that is not heard very often. At moments of high distraction or if they're about to chase something, the sound of the whistle will get into their head far more successfully, and from further away, than you calling with your voice would. Begin training the same way you would teach the dog its name. At home, with little distraction, blow the whistle and immediately give the dog a nice reward (possibly better than the normal dog kibble if you want to make the whistle extra special).

1. Once he has the general idea, you can move a bit further away from him: a few metres, one room to another room, and eventually outside.

2. You can also blow the whistle every time you feed him his dinner, to help build the association.

3. Remember to always give a good reward for returning to the whistle. If you don't have anything for him, then don't blow the whistle, or it will lose its effectiveness.

4. Similarly, remember not to overuse the whistle or it will lose its appeal. Practise recall on a walk with your voice more than the whistle. Two or three uses of the whistle per walk is plenty, to stop it getting boring for the dog. In the future, you will need it to be reliable in those tricky situations when all else has failed.

An extra note here – a clicker is *not* a whistle. A clicker is *not* to be used to get the attention of the dog, but to tell the dog that whatever it was doing at the moment of the click was great.

OTHER PROBLEM SOLVING

Never tell your dog off when they eventually return to you, no matter how angry you are. Returning to you should always be a positive experience. Try not to chase your dog – the closer you get, the further way they are likely to run. If it is safe to do so, try running away from your dog, and it is likely that they will follow you. Does your dog come back so far and then dance around, not letting you actually get them? Teach your dog that they get treats after you have touched the collar and clipped on the lead. Does your dog behave perfectly well until the end of the walk? They have probably learned that having the lead put on means the end of all the fun. Try putting the lead on for brief periods during the walk and then let them go off-lead again. If all else fails and you can't get near your dog to catch them, try lying down on the floor and playing dead. Many a dog has fallen for this and can't help going over to check if their owner is all right, at which point they are close enough for you to quickly reach up and take their collar.

TRAINING YOUR DOG TO NOT PULL ON THE LEAD

Sometimes our dogs do need to be on a lead, and this creates a fresh problem. The second basic thing every dog owner wants is for their dog to walk without pulling, because being dragged makes dog walking stressful and unpleasant. Training your dog to walk nicely needs to begin when you are not actually on a walk, when there is sufficient time to practise. If you try this on an actual walk, it won't work because it will take too long. Doing this at home also helps to teach the dog not to get too excited just because they have seen the lead.

STAGE 1

It takes two to pull, and many owners make the mistake of having a tight lead that forces the dog to pull against it. Start with your dog on either your right or left side (you choose, but be consistent!) and the lead not too long or short, but just loose enough. You are aiming to walk with your dog at your side, not in front of you. The lead should make a "J" shape. I find it helpful to have the dog on my left, and lead in my right hand, held around the position of my belly, so that my left hand is free to instruct the dog or give rewards.

STAGE 1: CORRECT POSITION FOR WALKING

STAGE 2

Many owners try to use food treats as a reward for walking on the lead, and these can be used on occasion. But the real reward should be you walking forwards, where the dog wants to go. When the dog is at your side, you walk. As soon as the dog creeps in front, and before he hits the end of the lead and has been rewarded for pulling for even a second by you moving forward, you must stop walking and quickly encourage him back to your side. To do this, encourage the dog with your voice and hand movements. I find taking a step backwards with the leg nearest to the dog helps create some space for the dog to move back towards me, and encourages him to follow me again. Do not just drag the dog backwards with the lead, or he isn't learning anything useful.

STAGE 3

From my observations, where many owners go wrong with this next bit is they wait too long when the dog is back at their side before walking forward again. Perhaps they take a few seconds to praise the dog, get a treat out of their pocket and give to him, or ask him to sit. What happens is a back and forth yo-yo effect. In order for the dog to learn that being in position by your side caused the walking forward, it has to happen quickly. So as soon as your dog is at your side again, praise and walk forward.

STAGE 4

Repeat the above many, many times. Sometimes you will be moving more backwards than forwards – a literal case of one step forwards, two steps back. What is most important is that your dog is learning that the only way it gets to move

STAGE 2: CORRECT THE DOG'S POSITION BY STEPPING BACK WITH THE FOOT NEAREST THE DOG AND ENCOURAGING HIM BACK INTO THAT SPACE BEFORE STEPPING FORWARD AGAIN

forward, which is where it wants to go, is if it is by your side and the lead is loose.

STAGE 5

Can you do the above exercises but without a lead? Or even with someone else holding the lead (if it is unsafe to be off-lead completely)? Once you do not have that back-up control temptation to tug on the lead to keep your dog in the correct position, you will find yourself working much harder with your voice and body language to keep your dog in the right place by your side, and their focus on you, not where they are going. In particular, you will have to respond quickly if they begin to wander away, which trains you in better dog handling skills.

ADDITIONAL EXERCISE

Another useful exercise for training loose-lead walking is the "red light – green light" method of Ian and Kelly Dunbar.[2] I have used this exercise during my training classes with great results, if the dogs are suitable and we have enough space. It does require a calm environment without much distraction.

1. With your dog on a lead and a pocket full of treats, simply stand there and look at them, but ignore what they are doing (and they will probably mess around a lot).
2. As soon as the dog sits or lies down, praise them and quickly give them a food reward (and enjoy the look of surprise).
3. As soon as the treat is delivered, take one large step in any direction (in a class situation, into a clear space), so that your dog moves with you and is now standing.

4. Once again, wait until your dog eventually sits or lies down before giving it another treat, and then take another large step.

5. Repeat many times, and you will begin to see the flashes of connection in the dog's mind that every time he sits, you give a treat and move. Soon, rather than frantically pulling and spinning and trying to get places, the dog will be standing next to you, expectantly looking up for the next reaction. At this point, take two steps, and then three, and before you know it, the dog is walking beautifully on the lead around the room. Bonus – as soon as you stop walking, they sit without being asked.

HEAD COLLARS AND HARNESSES

There are many harnesses and head collars on the market that can help with pulling, but they are not the magic wand that most owners expect. They still require you to train the dog, though they do make the training process easier. Avoid any head collar or harness that tightens up around the chest, head or legs when the dog pulls, as this is using aversion to make the dog uncomfortable (positive punishment and negative reinforcement). Likewise, avoid harnesses that connect to the lead only at the back, as these tend to make pulling worse as now the dog can put their full weight into it across the shoulders – remember harnesses like this are used for dogs to pull sleds, etc. Head collars and harnesses that connect at the front of the dog (i.e. underneath the muzzle or chest, not the back of the head or shoulders) enable you to gently guide the head or chest without needing force or inflicting pain. It is like having power steering, because the dog simply does not have the strength in his head that he has in his shoulders. Large,

TRAINING A DOG TO WEAR A HEAD COLLAR

1. Show your dog the head collar and feed him tasty treats from the same hand.
2. Hold the nose loop so that your dog puts his nose into it. Feed him treats as he does this (opposite.)
3. Let your dog put his nose through the nose loop for treats.
4. When he is enthusiastically putting his nose in, begin to move the straps as he is eating the treats.
5. Fasten the straps as he is eating treats and feed him more. Then undo the head collar, take it off and close the bag of treats.
6. Put the head collar on and reward.
7. Leave the head collar on for slightly longer. Give lots of rewards while it is on. All attention and treats stop when you take it off. Always ensure that you take the head collar off, rather than your dog removing it himself. If he tries to get it off, he must not be successful. Distract him with treats and make sure that he is tolerating or even enjoying having it on before you remove it.
8. Reduce the number of food treats.
9. Begin taking a few steps around the room with the dog in the head collar, attached to a lead. If at any point your dog tries to put his head down and start rubbing at the head collar, try pulling up gently using the lead to prevent him getting his head down and to keep his head still. As soon as he relaxes, release the pressure on the lead (so that he is rewarded for giving up – negative reinforcement), praise him with your voice and encourage

him to walk forwards again. Then perhaps consider going back a stage until he is more comfortable and less likely to react in this way.

10. Take him for a short walk.

These steps should take a few weeks to complete. At no point while wearing a head collar should the dog be walking with any tension or pressure on the lead. In fact, you should be able to hold the lead draped over your little finger. I see far too many dogs who have been taught to simply pull sideways into the head collar – which is even more damaging to their neck than walking on a collar and lead, and renders the head collar practically useless in preventing pulling.

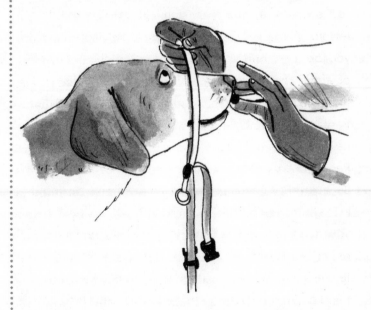

REWARDING YOUR DOG FOR PUTTING HIS NOSE IN THE HEAD COLLAR

powerful animals like horses are walked on head collars rather than collars or harnesses for this reason. With some harnesses, two attachment points can be used, one at the front and one at the back, for extra security and control.

Introducing a head collar must be a gradual process – if you simply slap it on the dog and go for a walk, you should expect them to freak out, and you may never be able to get near them with it again. Most dogs take a little time to get used to a head collar. Remember how he tried to get his collar off when he was a tiny puppy? This is similar.

I've only ever known two dogs that could not be trained to walk in a head collar because they absolutely hated it and freaked out, despite lots of training. One of them was my collie Ben, and in each case, we instead used a harness that attached to a lead at the front. Although technically, these are also using negative reinforcement to better control the dog, many dogs I have trained (including assistance dogs) have all worn head collars happily, because they have been slowly introduced. They actually got excited to see the head collars, as it signals that we are going for a walk. Although I have also trained my dogs not to pull on a collar and lead, I find walking my dogs on a head collar or harness useful for three reasons: 1) I have intermittent back pain, and any accidental pull or lunge during these times would be painful (interestingly enough, my dogs tend to walk best when my back is bad, probably because I am far less lenient at this time); 2) if you have a dog who is a bit of a klutz (as Ben was) and liable to walk himself somewhere into trouble, it is useful to be able to grab the harness and steer him out of a situation; 3) a harness, if used correctly, has less potential to damage the soft neck tissues like a collar would, in particular in a situation where the dog suddenly hits the end of the lead. All collars and leads, even basic flat collars (as well as ones that tighten) put pressure on the neck that can be damaging.[3] This means that it is critical that you train your dog not to pull on the lead, and you don't yank the dog's collar when walking. Using a harness (or for strong pullers, a head collar) can help you achieve this training.

REAL WALKING VERSUS PRACTICE WALKING

This all takes a lot of practice, and is quite difficult to integrate into real life. Sometimes you just need to get on and do the walk, and don't have time for a training session. A top tip is to have two kinds of leads for walking, one where your dog can pull a bit (but not drag you) and another lead for perfect walking. When you have time, you use the lead where no pulling is ever rewarded. When you need a quick walk, you use the other. Perhaps you teach your dog to never pull when it is on the head collar, but you are more relaxed on a collar and long lead. Perhaps your dog should never pull when on a collar and short lead, but it can pull a bit when wearing a harness. The choice is yours.

Over time, as your dog gets better at loose-lead walking, you will be able to spend less time on the pulling set-up and more time on the non-pulling set-up, until your dog is eventually walking well all the time. It can be helpful to bring both leads and practice the "nice walking" towards the end of the walk, when your dog is less excited.

It is also worth noting that just because one person has trained the dog to walk nicely for them, it doesn't mean that the dog will automatically walk nicely for everybody. Jasmyn rarely pulled me, because I didn't let her. However, if I quietly handed the lead to my mum mid-walk, she would immediately start to pull, because she knew she could. Jasmyn also taught Granny to feed her toast crusts, but that's another story.

A NOTE ON FLEXI- or EXTENDABLE LEADS

Many dog professionals hate these types of leads, for quite good reasons. Firstly, many dogs are killed or hurt when they run into the road on these leads, as people do not lock them at a short length, or they snap. I've had to screech my car to a halt when a dog ran in front of me on an extendable lead. Secondly, many dogs hurt their necks when they run and slam into the end of the extension. Thirdly, many people are hurt by being tangled or "rope-burned" by the string when a dog wraps it round them. When I worked as an assistance dog trainer, one dog was so notorious for this that the red mark round the back of your leg got nicknamed a "Dooley" after him. But as you know already, I am a "never say never" person, and there are some situations where perhaps for a small, calm, lightweight dog, with the lead attached to a harness (not the collar or head collar), they can be useful to give the dog a little more freedom on the walk in a safe green space. But, in general, they are unreliable and give little control over the dog in an emergency, so use a short fabric lead instead if you can – and use this book to teach your dog a reliable recall instead.

TRAINING YOUR DOG TO WAIT

The third and final doggy life skill is to wait. Although I also teach my dogs to stay (which we will cover later), they mean two very different things, and a "stay" is far less useful unless in emergency situations. "Stay" means do not move from that spot, I will come back to you; whereas "wait" means "Hang on a second ... OK, now carry on." "Wait" has so many uses: wait while I put your lead on; wait before we get out of the car; wait – don't eat that yet; wait while I throw ... now go fetch it! Most importantly, teaching a dog to wait helps it to learn a bit of self-control. As we all know, good things come to those who wait.

NOTE ON RELEASE WORD: If we have a word which means "do not move", we must have a word which means "now you can move". This is a release word. I use the word "OK", but you can use "finish", "go play", etc. This should be taught along with the "wait" command. It means:

"You are off your lead, now you can run";

"I have opened the door, now we can walk through it";

"I have opened the car door and am holding your lead, now you can come out," etc.

In other words, it means "OK, you can now move." He must not get the reward of moving until you have told him to move.

Do not use "good boy/girl" as your release command, as you want to be able to use that to praise the dog for doing the right thing, without it serving as permission to stop whatever he was doing.

STAGE 1

1. Have your dog on a short lead and kneel down beside him. Shorten the lead so that you are holding it near the collar and can restrain the dog from moving forwards.
2. Show him a nice treat in your hand and ask him to wait (I use a hand signal of a closed fist with my index finger

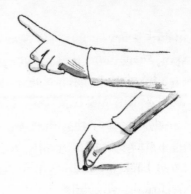

STAGE 2: TEACHING A WAIT

pointing upwards – see below). Place the treat on the floor within view, at an arm's length away.

3. Restrain him there with the lead as he tries to get to the treat. As soon as he relaxes and stops pulling, praise him, give your release command and let him go get the treat.

4. As you keep repeating the exercise, you will notice the dog getting quicker to relax. He may even sit, which you can praise. Now wait a few moments, then release your dog using the release word.

Remember to be consistent with the release command as you do not want the dog to release unless told. This could be a life or death situation, such as at a roadside.

STAGE 2

As the dog gets better, you can ask for a sit first (see Chapter 10) and hold the sit until released. Then progress to no lead and holding the sit until released. Remember, you will need to be ninja-quick and get to the treat first and take it away if the dog tries to get it before being released. It is imperative that the dog does not get to the treat unless he's told he can!

STAGE 3

Next teach him to wait for his dinner. Once the dog understands what "wait" means, try using it when putting his food down. Ask him to sit and wait. As you move the bowl slowly towards the floor, he should still be in the sit position. If he leans forward or gets up, bring your hand and the bowl back up and ask him to sit and wait and start again. Eventually you will be able to put the bowl down gently on the floor and move away back up to standing position. Be ready to grab the bowl if the dog tries to get it. Praise the dog gently (but don't distract him from his sit) and give your release command to go get the food. The whole process may take a few minutes but hey, it could mean no more being mugged at mealtimes!

STAGE 4

Next teach him to wait at doorways. It is important that your dog does not charge through doorways ahead of you, both for safety reasons and because we want our dog to be enjoyable to be around rather than rudely bowling us out of the way. The wait command and subsequent release can be used going through doorways; the same principles apply. By now, your dog should have a good understanding of "sit" and "wait". If he breaks the wait, shut the door, put him back in the same place, and start again. Remember to praise gently for waiting. You can also give a treat if you want, but the mere action of going through the door to go on a walk is rewarding to the dog. The release command does not mean a free-for-all bomb through the doorway either. You should walk calmly together. Go back and repeat until this is accomplished, rewarding and praising for steady walking.

The same principles apply for jumping out of the car – make your dog wait in the car to have the lead clipped on and until you say it is safe to get out. If he breaks the wait, put him back in, shut the car boot/door, and start again.

The wait command can also be turned into impressive tricks. Will your dog wait with a treat on his paw until you tell him it's OK to eat it? Can you even balance the treat on his nose?

MAKING SURE EVERYONE ENJOYS YOUR DOG BEING A DOG

I believe that if you do no other training with your dog, but do master these three things, your pleasure and satisfaction with your dog will be greatly enhanced. So will those of others impacted by your dog, such as other dog walkers. By teaching your dog to come when it is called, you are able to enjoy letting your dog off the lead without disturbing or annoying anyone else. By teaching your dog to walk on a loose lead and not pull, you will find walking your dog (to where it can be off-lead) pleasurable. By teaching your dog the patience to wait temporarily when asked, you will find your dog much easier to be around and to take places. If you are able to enjoy walking your dog and spending time with him, your dog will also have a happier life and be able to spend more time out and about.

CHAPTER 9

MAKING MORE TIME FOR WALKIES

So now we have a dog that comes when called and doesn't drag us around, but we might still not be walking it as much as we want to. Finding a way to increase the amount of walking you do with your dog is a sure-fire way to improve both your physical and mental health, so long as the dog is well trained. If you learn nothing else from this book, at least take your dog for a walk. Like right now. Go!

Did you enjoy yourselves? Or did you think, "I'm too busy right now, I'll do it later"?

There is a big difference between knowing we should do something and actually doing it. The biggest issue we face with motivation for dog walking is time. Our modern lives are incredibly busy, and it can be difficult to fit in healthy activities. I realize that it may sound simple to just make more time for dog walking, but it's not – this is the part of the book that will provide you with the secrets of developing an intention to walk, and then translating that intention into real action.

WHY WON'T JUST TELLING PEOPLE TO WALK THEIR DOGS WORK?

Whenever the intention is to change a person's behaviour, there is an assumption that "more education" is the solution to the problem. Therefore, what happens if we tell people that they should walk their dogs more? Does it work? You would think so, especially if we used the incentivizing message that it was good for the dog's health. Professor Ryan Rhodes is a Canadian physical activity behaviour researcher, and he tried just that.[1] Over 12 weeks, inactive dog owners in both the control group and intervention group increased their physical activity (measured by pedometers and questionnaires), but step counts were a bit higher in the intervention group – promising, but not convincing. Since Ryan's original study, together we have performed a review of 13 different intervention studies we found that tried to use dogs to increase people's physical activity.[2] They show that human–canine interaction can motivate physical activity behaviour, but how effective the intervention is depends on how it is designed. Dog walking doesn't appear to be a done deal – it is also influenced by many of the barriers and motivators to physical activity we already know about.

WHY DO SOME PEOPLE WALK THEIR DOGS LOTS, OTHERS LESS, AND OTHERS NOT AT ALL?

According to my interviews with UK dog owners, there is an unwritten general rule that dogs should be taken for a walk every day.[3] If you are reading this from elsewhere and find that surprising, perhaps the societal views in your location are different, illustrating how cultural beliefs can influence what is expected and seen as "fact". However, no matter where you live, I don't think the problem is that people don't think dogs need walking. No person, no matter where they are from, has ever said to me, "You mean to tell me that dogs should be taken for a walk? I never knew that, thanks!" This may help to explain why Ryan's results were not as dramatic as he had hoped.

As we discovered in Chapter 4, many dog owners don't walk their dog every day (even in the UK), or at least don't always do the walking with their dog themselves (maybe a different person in the household walks the dog). Many of the owners who told me that dogs should be walked every day because it is required for their physical and mental well-being, didn't walk their dog every day, and gave me many reasons why, in their particular case, this was the appropriate choice.[4] Many of these arguments were surprisingly based on the principle that it was best for the dog not to be walked. I will explore some of the common justifications (or perhaps excuses) in this chapter, and how you can tackle them.

Before I started my research into why people walk their dogs, I reviewed the already published studies describing what factors are associated with dog-walking behaviour.[5] Although it is tempting to place responsibility for any health behaviour squarely on the shoulders of the individual and the choices they make – such as how much alcohol they consume or what food they choose to eat – in reality, we know that our behaviour is constrained and influenced at many levels, from individual choices up to policies applied to the whole of society. For example, we can only buy food that is available to us in the shops and at a price we can afford; it's incredibly tempting to join in for a drink if all your friends are having one and you've had a stressful week at work.

For our dog-walking behaviour, we are potentially influenced by policy (where dogs are allowed to be walked off-lead), our physical environment (how close we live to green spaces), our social environment (whether our friends and family also have and walk dogs), and our personal demographics. On the latter point, even more complicated by this particular scenario is that we are not only influenced by who we are (how old we are, whether we work), but also by who our dogs are (how old they are, their size/breed). As people and dogs come together, we intertwine and form a unique relationship. Interestingly, all the evidence pointed to this being the most important factor in determining dog-walking behaviour, in terms of both the amount of evidence (number of studies with significant findings) and the effect size (how big a difference it makes).

There is something about the relationship some people have with their dogs that provides a strong sense of responsibility (or "obligation", as Ryan called it, or "motivation and social support", as Dr Hayley Christian found) for going for a walk.[6] I think these are all slightly different ways of measuring aspects of our relationship with our dogs. As I thought about this, I wondered if it was a chicken or egg situation – which came first? Does a stronger relationship mean that we want to walk our dogs more, or is a strong relationship developed through spending time walking with your dog? My subsequent interviews with dog owners suggests it's a bit of both. The interactions and relationships we have with our dogs, if they are good ones, make us feel responsible for caring for them and meeting their needs. As some of the children I interviewed put it (children give the best quotes, they say what they really mean):

"Because we love them so much and if we didn't really love them then we wouldn't take them to the park, but because we love them, we've got to take them to the park."

"I would say if you want to do the right thing for your dog, make your dog happy, then take it for a walk, make it all nice and happy."

Through spending time walking with them, this relationship gets strengthened further, and we feel even more responsible for making them happy and providing them with a good life:

"Well, there is responsibility for another animal ... you are responsible for everything to do with his welfare, his upbringing, his training, *everything* ... He gets lots of exercise, a minimum of two hours a day. He gets nice food. Well, he's never complained [laughter]. Yeah, he is treated well, he's not mistreated, he's NEVER been hit ... I think he has got a good life."

But here's the crunch bit – seeing our dogs happy makes us happy, and that's the real motivator:

"I thoroughly enjoy it because I think he enjoys it and I love the thought of him being happy, so to know that he is out somewhere new and he is enjoying himself and that he's allowed to sniff and he is bouncing around and you can tell that he is excited, I love that."

Although we say dog walking is "for the dog", it actually provides us with huge benefits ourselves, such as relaxation and stress relief (as we discussed in Chapter 5). Walking with a dog is even better than going for a walk by yourself:

> "It just feels special when they're there with you. It makes a good walk an excellent walk. It's that little bit more when you've got an animal by the side of you."

Thus dog walking can make us feel good, which is the secret to its success, but also its demise. Ryan's subsequent research agrees with my interview findings, that dog walking is largely "intrinsically motivated".[7] It's not so much something we do because of potential external influences (extrinsic motivation) – because our dog pressures us, in order to get a pat on the back from other people, or to stave off negative feelings such as guilt. It's something we do mostly because we just enjoy doing it.

The problem with dog walking being so intrinsically motivated is that if dog walking isn't pleasurable to do, we aren't motivated to do it. And it only makes us feel good if it's done right. One of the major issues contributing to us not enjoying walks is if the dog has behaviour problems, or not even problems as such, but is just a bit of hard work to walk. This is why in the previous chapter I taught you how to address the two most common behavioural/training issues leading to displeasure on a walk: not coming when called, and pulling on the lead. If behaviour problems are even more serious than that, owners can find it easy to justify to themselves that it is "best for the dog" not to be walked:

> "He was the main reason why we started to realize that dogs don't always need to be walked every day. For him, it was too stressful to go on a walk every day."

I hear a lot from owners (and behaviour counsellors) that for some dogs it is better not to be walked. I am sympathetic to their point, but I'm not sure I completely agree. For sure, for dogs in significant behavioural cases, when they are very nervous, it is not a good idea to overwhelm them with a busy place full of people and dogs that completely freak them out – that's not enjoyable for the owner or the dog (or helpful to the dog's recovery). However, not walking the dog at all doesn't solve the problem. It should be possible to find some way of walking the dog, perhaps in a quiet area away from other people and their dogs, so that the dog (and owner) can still glean the mental and physical benefits of exercising in nature. Avoiding walking the dog also does not do anything to address the behavioural problem of the dog, which isn't going to go away on its own.

ACTION POINT

If you want to develop a sense of responsibility to walk your dog more, then find other ways of strengthening the relationship that you have with your dog. This could include spending more time with your dog: snuggling on the couch, teaching him tricks, playing with toys, or taking him to training classes. And it doesn't have to be boring training. There are now many different "dog sports" options to try, including agility, obedience, heelwork-to-music, scentwork, canine parkour, hoopers, rally, frisbee – the list is endless. However, this approach is not going to be effective unless you are actually enjoying the time you spend with your dog. Thus, you need to address any training issues or behavioural problems as a priority. Some of these we have covered already, but more advice regarding how to approach dealing with behavioural issues awaits in Chapter 13.

Ryan's research did find another strong predictor of dog walking, independent of whether the owner found dog walking intrinsically motivating, and that's the size and breed of the dog. In our review, we found that a number of studies demonstrated that smaller dogs were less likely to be walked than larger ones. Some of the logic behind a perception that small dogs need less exercise was explained to me by my research participant:

> "Well, the size of a dog, take your Roxie. The amount of steps that she's got to take per metre have got to be far greater than what Ralph [an Alaskan malamute] does. So I wouldn't expect her to need as much exercise as him."

However, research into the association between size and exercise occurs across an average of a population, and I know plenty of owners of small dogs that perceive that their dog needs a walk every day, and do as such. Anyone who has spent a lot of time around different dog breeds also knows that size is a bit misleading. When one of my research studies suggested that owners of larger dogs are more motivated to walk than owners of smaller dogs, one newspaper article recommended we should all get Great Danes (and I held my head in my hands in disappointment). In fact, many large breeds are perceived as slow and lazy, and small breeds as highly active. In order to investigate this, one of my undergraduate students analyzed a dataset of over 12,000 survey responses on breed and exercise, given by pedigree dog owners who had watched a TV programme about dogs featuring my colleague Professor Alex German.[8] Granted, even with such a large dataset, the numbers of each breed were low for making statistical comparisons, but I noted how there were large dogs in both the most and least exercised list – even of the same type: hounds.

Following this study, I purposefully interviewed owners of particular breeds in order to dig deeper, and discovered carefully constructed views about the exercise needs of each breed – and characterizations of them as being slow or lazy, or difficult to walk in other ways, despite them being of similar sizes to other breeds perceived as highly active. So rather than there being a blanket view that small dogs need less exercise, there are particular small (and some larger) breeds where preconceptions about exercise needs may need addressing. If you perceive your dog to need high amounts of exercise, you are more likely to walk it.

So how much exercise does each breed actually need? There are no evidence-based guidelines on this; any recommendations are simply based on the views of people who own the breed. It is worth noting that The Kennel Club UK recommends 30 minutes each day for even the smallest breed, so that should be the bare minimum you are aiming for. For some breeds, they recommend over two hours. Personally, I feel that most dogs are satisfied if given a 45-minute to one-hour walk each day, but do enjoy more when possible.

Aspects of the dog that can affect our motivation to walk them also include the age of the dog. As Jasmyn aged, I did notice her slow down, but she still seemed to enjoy a daily walk, just at a more leisurely pace. Ben, on the other hand, did not slow down; he was the type of dog that would run and run until someone physically stopped him. Is your dog genuinely lazy or old, or are you making excuses for yourself? I'm not judging here; I have proof that I've done it myself, as evidenced in the diary entry I wrote while doing my research:

"I was working from home and the day flew by, and it was early afternoon and I still hadn't had time to take the dogs out. I found myself thinking about someone I had recently interviewed and how they justified giving the dogs a bit of play time instead of a walk, if they don't have time to take them out. So I took Roxie out in the garden and played fetch for a few minutes. Jasmyn looked tired out so I let her sleep. I thought, 'She's getting old now and will be OK without a walk today, I will give her a rest.' ... As I was throwing the ball again for Roxie in the garden I suddenly thought, 'WHAT AM I DOING?' It was like a wake-up moment ... I really shouldn't NOT walk them. This project makes me feel hugely guilty when I notice myself constructing justifications as to why I don't need to walk them, because it fits with my own needs that day. So I put their leads on and took them out. Although we only went around the block. Better than nothing, but I was really running late!"

Very young dogs also need to have their exercise carefully managed, so that they don't overdo it, get too tired and potentially damage their developing joints. There is limited scientific evidence on this topic too (likely because it varies by breed), but a good rule of thumb is a ratio of five minutes of exercise per month of age (up to twice a day) until the puppy is fully grown.[9] This would mean two 15-minute walks when three months old, two 20-minute walks when four months old, etc. Calculate what your puppy can do.

ACTION POINT

Challenge your perceptions about what exercise your dog requires. Some people think small, old or lazy dogs don't need as many walks as bigger or younger ones. It is actually possible for a chihuahua to climb a mountain or run five miles (I have proof!). Are you constructing excuses for yourself so that you don't have to walk them? Ask your vet for advice as to what your dog could realistically do, and if in doubt, start with a short walk and build gradually.

The quote from my diary above gets to the root of the third big barrier or motivator for dog walking – how we can fit that in with the rest of our lives, and our personal needs, even if we enjoy dog walking and think our dog needs it (as I certainly do). Much of the rest of this chapter will address this issue, however, one quite genuine barrier should be mentioned first. As well as thinking our dogs are possibly incapable and don't need walks, we also have our own health issues. We all have the odd days where we have a short-lived acute health problem preventing us from leaving the house (or bathroom), but not a lot can be done about that. I would say that the only other time I really don't get out on a dog walk for possibly a few days is if my back is bad. The irony is that movement is recommended for chronic pain, but I know it is easier said than done.

We have already discussed the need to challenge our dogs' capabilities. Likewise, you may think that you are too unfit or have health needs that mean you can't walk yourself, or at least not far. However, it is always worth revisiting this perception, bearing in mind that physical health needs do not appear to be associated with whether owners are able to regularly walk their dog when we look across data averages. What I mean is, many people with poorer health, including those conditions that affect walking, are not actually prevented from walking their dogs. Walking is arguably the easiest, most accessible, and most gentle form of exercise. Granted, you may have to think carefully about where best to walk (my interview participants suggest avoiding uneven surfaces and finding benches to rest upon) and perhaps use aids (walking sticks, wheelchairs, walking with another person for support), but it may be that you are capable of more walking than you first think. Always seek medical advice before starting a new exercise regime.

ACTION POINT

Are your own health problems and fitness stopping you from walking your dog? Again, seek medical advice if need be, but in most cases you can start going for short slow walks, and build up distance and speed from there. If you are already walking, can you aim to walk for longer or faster?

It appears that our perceptions and personal beliefs about our ability are more likely to be associated with dog walking (or motivation to dog walk) than any other specific factor about us as owners, such as our age, work schedules, etc. Basically, if we have the right frame of mind, we can get around many barriers. However, it does help if we have friends and family that are supportive and encouraging of walking, and it doesn't help if we allow other household members (such as spouses or children) to walk the dog instead of us.[10]

One more recent development that is probably not accounted for in past research is the rising popularity of paying somebody else to walk your dog instead, or sending your dog to "doggy day care" when you are out at work. This is a striking change noticed by my behaviourist colleague Erica Peachey. She tells me:

> "I've noticed that when I ask my training clients if they have a dog walker, they sound almost apologetic if they tell me that they don't, like there is a perception that it is part of being a good dog owner to have a dog walker or send your dog to day care. I don't necessarily agree. Should you get a dog if you haven't got time to walk them, although they can be a godsend in particular circumstances. Is it always a good thing if they come back from day care so tired, given that dogs naturally spend a lot of their day sleeping? Perhaps they find the experience of being around all these other dogs stressful? I also find that using these services can sometimes undermine a dog owner's confidence and ability with their own dog – they struggle to know what their dog likes or dislikes, or how it reacts to things, because they only walk at weekends."

Like Erica, the excessive use of dog walkers and care services worries me, but from my perspective it's because it's such a waste of a potential exercise opportunity and the owners are literally missing out on the best part of having a dog. However, these services can also allow people who wouldn't otherwise be able to own a dog – because they work in an office away from home, or are dealing with a health issue – to benefit from the other aspects of having dogs in their lives, or transition through a difficult period in which otherwise they may feel the need to

give the dog up. Used carefully, they can be a fantastic comple-
ment to the care that the owner gives, but make sure that they
aren't being used to replace it. An owner must also carefully
choose an appropriate service that does not undermine their
training principles, as we discussed previously.

ACTION POINT

Are you allowing someone else to walk the dog instead of
you? Can you make it your responsibility to walk the dog
instead? Even if your dog has already had a walk that day
by someone else, can you take him for another walk that you
both would benefit from? Or can you arrange schedules so
that you all walk the dog together?

One important factor that we perhaps have less control
over is whether we live near to suitable places to walk our
dogs. In one of Dr Hayley Christian's studies, dog owners
were more likely to walk their dogs regularly if they lived
within 1.6 kilometres of a park with dog-supportive features.[11]
But what does a good dog-walking space look like? Firstly, it
needs to allow dogs off-lead, as we know that seeing dogs
running around enjoying themselves is a major draw of the
dog-walking experience , as research participants tell me:[12]

> "Well, I don't think walking on the pavement is very
> interesting for them. If I see a dog walking on the
> pavement, I think it must be really boring, being on
> your lead."

> "I look forward to going to the park and just seeing
> the dogs going running."

However, care needs to be taken to ensure that people involved in planning local amenities do not use this as an excuse to fence in a small "off-lead" dog park, for which there are many concerns about encouraging conflict between dogs, and inactivity of their owners. According to my research participants, dog-walking locations perceived as "good" were large and had interesting scenery and circular routes to avoid repetition. These locations also had to be "dog-friendly", mainly encompassing aspects of safety, such as suitable walking surfaces, and being able to avoid poisons, livestock and vehicles, so that owners could relax and enjoy themselves. I suspect that, when the choice is open to them, dog owners will choose to live near good dog-walking spaces, as it certainly factors into my decision when choosing a house. However, many people will have less control over where they live, and policy-makers and planners must consider the provision of local green spaces to encourage physical activity of both dog owners (likely a quarter to a half of their constituents) and people without a dog.

ACTION POINT

You may benefit from making an effort to find new and pleasant spaces to walk your dog in. Or maybe you are like my father-in-law, happy to walk exactly the same route, time and time again. I have a suggestion for everyone: next time you take your dog for a walk, do the route backwards, just for a bit of fun. No, don't physically walk backwards, but reverse the route. How odd this feels proves that dogs and their owners are both creatures of habit. Who knows – a simple change like this may help you get out of your dog-walking rut and inspire some new adventures!

HOW TO CHANGE YOUR
DOG-WALKING BEHAVIOUR

There are many different theories to explain why people behave the way they do (and I am not going to explain them all here), but a central idea is that before a behaviour (such as going for a walk) can happen, there first needs to be an intention to perform that behaviour. This intention is influenced by a number of different attitudes (our personal beliefs about the benefits or drawbacks of performing the behaviour), subjective norms (what the people around us think and how much we care what they think) and perceived behavioural control (our perception of how difficult the activity is to do in practice). [13] In behavioural surveys, having an intention to walk your dog regularly is an important predictor of actually walking your dog.[14] During the first half of this chapter, we have addressed some of the perceptions and beliefs that I have commonly found would lead us to form an intention to walk our dog on a regular basis. These include our ability, our dog's ability, and whether we enjoy sharing this aspect of our relationship with our dogs.

ACTION POINT

Solidify your intention to walk by writing a list of why you want to walk your dog more. How will your dog benefit from more walks? How will it benefit him physically? Are you expecting any behavioural benefits? What will be the benefits to you? Do you wish to be physically fitter? Will it help with your stress relief and mental well-being? What is currently stopping you from walking more?

Making a list reminds yourself of why you're doing this and what you want to achieve. Put it up in a place where you will see it each day as a reminder, such as on your fridge or kitchen pinboard.

Reasons I want to walk my dog more	
Benefits my dog will gain from more walking	Benefits I will gain from more walking

TRANSLATING INTENTION INTO ACTION

Intention isn't the end of the story, though. The majority of dog owners have positive intentions to walk more, yet almost half fail to meet those intentions.[15] We have already seen how feelings of self-efficacy and perceived behavioural control are areas which may scupper our best intentions to walk, in that we aren't confident that we can actually achieve it, or feel that other things are going to crop up and get in our way. Intention is known to predict a considerable amount of physical activity variability, but it does not explain all physical activity undertaken, known as the intention–behaviour gap. Action control then describes people's ability to further translate intention into action.

Professor Ryan Rhodes found that what distinguished those dog owners without intention to walk, those with intention to walk but it didn't happen, and those with intention and did walk, was the way that they thought about their dog walking and also practised it. Ryan suggested that interventions should focus on affective judgements (e.g. finding more enjoyable places to walk), behavioural regulation (e.g. setting a concrete plan), habit (e.g. making routines and cues) and identity formation (e.g. affirmations of commitment). These interventions may help overcome difficulties with translating these intentions into action. My qualitative interviews with dog owners also echoed similar messages about what the difference is between those who dog walk a lot and those who don't. Many of these aspects are also not peculiar to dog walking, but well-known techniques to be used in promoting any type of physical activity.[16] Thankfully, this means that even though we may feel we aren't in great control of our ability to dog walk, there is a lot we know we can do to address this and make it happen.

SCHEDULE YOUR WALKS

Probably the most important thing you can do with your good intention to dog walk is plan exactly when you are going to enact it. This means taking a few moments to look at your schedule for the next few days and decide who is going to walk the dogs, exactly when, and for how long. Coordinating the hectic schedules of two full-time professionals and a school-age child takes precision, and in our house, if it isn't in the diary, it probably isn't happening. Schedule your walks as appointments in your calendar, just like you would for a work meeting. We are far more likely to follow through on commitments that we have made in advance, even if just with ourselves. I know that this is the secret to controlling my personal behaviour – I could do yoga practice by myself at home, anytime – but I never do. If I am going to achieve my goal of doing yoga, I need to book on to an actual class, and make a commitment with myself that Tuesday at 7:30 p.m., I am doing yoga, and it is booked into my diary. Of course, it also helps that I have paid for it in advance.

When thinking about the best way to plan your dog-walking schedule, be easy on yourself. Remember to set small, achievable goals for the first few weeks and build difficulty slowly. Avoid planning your walks too close to other daily events so you have a buffer (for instance, my work tasks always run over). Don't forget to be realistic – if you're not a morning person, don't plan your walks for 6 a.m. Think about potential pitfalls in advance. For instance, if you know that nothing gets done when your kids get home from school, don't tell yourself it's going to be your walk time.

If you want to commit to getting up first thing and getting it done before the day starts, make the process as easy as possible on yourself by preparing the night before. It's not fun stumbling round the room in the dark trying not to wake anyone while you look for your dog-walking clothes. Some effective runners and swimmers actually sleep in their exercise clothes so that they don't have to get changed in the morning! Given the typically muddy state of my dog-walking trousers, I won't be trying that personally, but I do make sure they are laid on a chair, ready to go.

At this point, you may be looking at your calendar or diary with dismay and wondering how this is ever going to work. You may not even use a calendar or diary. In the latter case, maybe you have bigger life scheduling or productivity issues than just dog walking, and you need to start using one. Alternatively, you may be retired and have all the time in the world to walk your dog, but it still doesn't happen much (funny how that often happens). If you find yourself in these situations, it can be helpful to go back a step further. For one to two weeks, write down how often and how long you walked your dog, how you felt when you did it, what else you had on that day, and why you did, or didn't, walk. Where do you spend time in your day that could be spent walking instead?

WEEKLY DOG-WALKING DIARY

Day	Dog walk – Where? When? How long?	How it felt	Other commitments and activities I did that day	Why I did or didn't walk
Mon				
Tue				
Wed				
Thur				
Fri				
Sat				
Sun				

Don't forget to continually review your progress. Plans are just that – plans – and they can be changed as you need. Don't let not feeling well one day or a surprise meeting mean you lose one of your walks for the week; schedule it in elsewhere instead. If you are struggling to meet your targets, try splitting your walking time to better fit your routine. If you only have time for an additional 10-minute walk on a busy day, that's OK. Something is always better than nothing.

ACTION POINT

Take a few minutes to ask yourself when would be the best time in your daily schedule to walk the dog, where you will do the walking, and commit to it in your calendar by logging it as an appointment. If you struggle to get up early to walk, lay out your dog-walking clothes the night before so it requires less effort in the morning.

CREATING A HABIT

During my interviews with dog owners I noticed that the most successful dog walkers were those who have a routine – they walk at the same time every day and have it scheduled in. Ryan also found that habit formation, or measures of "automaticity", was a significant predictor of dog-walking behaviour in his quantitative surveys. Humans are creatures of habit. Luckily, so are dogs. Once you begin walking with your dog, you may soon find your dog is in control of your walking timetable.

When I was researching this behaviour, at first I was alarmed at how badly trained some of the dogs appeared to be. "I would never let my dogs pester me to be taken for a walk," I thought smugly. But I began to rethink this, as a pestering dog who "knows the time" is incredibly difficult to say no to. If you want more motivation for walking, let your dog learn the schedule and make sure to reward any signs of excitement with praise and then with actually going for a walk. When I rescued Brie, she absolutely loved her newly found freedom on walks, and would yodel at the slightest hint we might be going (dogs can quickly learn which are our dog-walking shoes or coat). To my initial shame, I deliberately continued to let her do this, but seeing the joy on her face has been a huge motivator for me. You may even want to consider rewarding specific behaviours, such as fetching the lead and looking at you with big eyes, to help your dog motivate you.

Habits are important for maintaining and sustaining human behaviours, as activities that are formed into a regular habit require much less conscious thought and preparation. Therefore, the more you can make dog walking a habit, the less effort it will take to do it because it will become an automatic part of your life. Of course, once habits are formed, they are also hard to break, so are more resistant to being derailed by competing activities or interruptions. Back to my yoga analogy, it is now my instilled habit that every Tuesday at 7:30 p.m., and Friday at 6 p.m., I do a yoga class. If a request tries to interrupt that schedule, I get quite grumpy – I am resistant to changing it – and I am now much better at following through with my intention and actually practising yoga.

MAKE A COMMITMENT TO OTHERS

As we discovered earlier, dog walking is inherently a social activity, whether you want to talk to people or not. Perhaps you are a social butterfly, and meeting with others in a group or even just one person will help you to be more motivated to take more frequent, or longer, walks. Although I am pretty good at going dog walking, I am less motivated to do any other physical activity, which I find far harder work than a nice, relaxing dog walk. Debra is my exercise buddy; for many years, we went to an aerobics class together each week. In the past year, due to the pandemic, we've started running (cliché, I know). I've had to pull out every trick in the book in order to spark any motivation to run, and Debra is my trump card. When she can't be bothered (which happens quite frequently) she knows I will be quite annoyed with her if she bails on me, and doesn't want to let me down. I think she starts at least 50 per cent of our run conversations with "I nearly didn't come today." For me, the power of running with Debra is less about having made a commitment, (if I tell myself I'm running that day, that personal commitment is usually enough), but more about distraction – it's far less painful an experience if I have someone else to chat to and share it with.

Public accountability can also work for motivation. You're more likely to achieve your overall walking goal if you have a commitment to work towards, such as a walk to raise money for charity – even better if you've told everyone on Facebook that you are doing it. For Debra, running is all about signing up for medals and entering events, whereas I couldn't care less about getting a medal (I am getting the feeling that Debra is more extrinsically motivated than me). Debra signs up for the medals, and then I have to run with her – we all win in our own way.

ACTION POINT

Science shows that we are less likely to back down on our promises if others know we've made them. Find a friend or family member to walk with you when you can for an "accountability buddy". Perhaps you would enjoy joining a larger dog-walking group for a regular group walk? Can you set yourself a challenge to walk a certain distance and make money for charity?

While we are on the subject of walking with others, I thought it would be pertinent to address the issue of dog walking with other household members of the small and whiny variety. It can feel like it would be a lot easier to walk without the kids. However, family dog walks benefit every-one in the household and can be enjoyable in their own way. It's probably the safest time for dogs and young children to spend interacting, and can really help build their positive relationship. My top tip for making family dog walks happen as much as possible is to once again be prepared. If your child is in a buggy or pram, buy one with suitable wheels for off-roading (not tyres that puncture), and seek dog-walking routes that have flat, mostly dry paths. Through the advice given in Chapter 8, train your dog to walk nicely beside the pram so that you can more easily manage both the dogs and the kids. You may find investing in a child carrier or sling a helpful alternative to trying to push a pram through mud. For the older children, taking a bike or scooter with you can help fend off cries of "My legs are tired!" (although you may need to be prepared to drag or carry it yourself for part of the walk). Suit the kids up in waterproofs and wellies so that you don't have to worry about how many puddles they jump in (and you can hose them down alongside the dog on your return). If you are going to be out for any significant amount of time – walks with kids are inherently slow and always take ages – remember to take snacks and a drink. Plan activities that can be done on a walk: what can you see, build and find? Our discovery of geocaching, where people leave clues to treasure that can be found all over the world in everyday places, has transformed many a humble dog walk into a brilliant family adventure.

Above all, try to enjoy dog walking with dogs and kids for the chaos that it is – Gruffalo hunts, grazed knees, and muddy hands and paws. In one of the places we walk only a few times a year, there is a fallen log on which we often take a photo of Brandon and the dogs sitting. As time passes, he grows taller and the dogs greyer, and it is a lovely reminder of that quality time we spent together. Who needs professional family photo shoots when you have dog walks?

OTHER WALKING TIPS

When it comes to going for a walk, most dog owners are less affected by weather than people without a dog,[17] probably because we feel that most of our dogs want walking whatever the weather. Many of my research participants have used the phrase "There's no such thing as bad weather, just bad clothing!" However, some claim that their dogs don't like the bad weather either. Remembering Roxie's face today at 7 a.m. in the dark and rain, I would probably agree that for a few dogs this is probably a genuine observation (in contrast, Brie seems oblivious). That said, just because my child would prefer to eat pizza every day doesn't mean that it's good for him. Whether your dog gets walked in the rain is primarily your choice, as their owner.

ACTION POINT

Is it your dog or is it really you who doesn't like the rain? Invest in some warm and waterproof clothing for you, and for your dog if needed. A head torch for you and lights for your dog's collar and harness are other excellent investments to help make walking in dark winters feasible and safer.

During my research interviews with dog owners, I discovered that there were actually different types of dog walks: one type that I termed "functional" and one "recreational".[18] When conducting qualitative research, which requires interpreting themes from long transcripts of conversations you have had with participants, it is challenging to know whether you are projecting your own feelings and beliefs on to your findings or really letting the data speak for itself. When my study was under review for publication, and vet researcher Dr Zoe Belshaw published her own study of owners of dogs with osteoarthritis, claiming that there were two types of dog walks, "functional" and "leisure".[19] We were overjoyed that two people who had never met had independently come to the same conclusion from studying separate datasets, solidifying our interpretation that this was a real thing and we hadn't just made it up. So what is the difference between a functional and a recreational walk?

Functional walks are those walks that are driven by feelings of guilt to provide the dog with a convenient form of exercise and a means to toilet outside, but are not much fun for the owner. Think dark winter mornings before work, on a lead, perhaps just round the local streets or park. In contrast, recreational walks provide significant owner stress-relief and are longer, typically during pleasant weather and at weekends, in less urban environments, and involve more members of the household. One of my interviewees explained:

> "That was a completely different walk than the ones I do in the woods. The ones that I do when I've got to go to work or just to the common or just to the fields to me are literally just for Bear. Even though I enjoy it because I'm walking with Bear and I'm out, they were just functional walks. The walks that I did in the woods were for his benefit and for my benefit and they were a completely different feeling for me than the functional ones."

It is worth engaging in a moment of self-reflection as to what dog walking means to you and whether your current walks are mainly functional or recreational. What type of dog walk would you prefer to do? What type of dog walk may be easiest to fit into your lifestyle at this point, if you were to start walking more? Thinking through these points may inspire you to aim your goals at a particular dog-walking strategy that you are most likely to benefit from. For example, if you are in dire need of some stress relief and quality time with your pet, then you should aim for increasing your number of recreational walks. How can you make available significant chunks of time, flexible around the weather forecast, perhaps driving to a really beautiful location? In contrast, if your schedule is jammed full, how can you leverage the canine need to burn off a bit of energy and squeeze in an extra functional walk each day, perhaps getting up half an hour earlier so that you can take a stroll together before the working day? I adjust my dog-walking plans during the winter months in order to achieve more recreational walks. If I only walked before and after the working day, dog walks would be dark and pretty miserable. Instead, I try to work from home some days and leave myself a proper lunch break, so that we can walk in daylight.

ACTION POINT

Are your dog walks functional or recreational? What would you like them to be? If your walks are mostly functional, can you fit extra recreational walks into your life? If you rarely dog walk at all, can you begin with some short, functional ones for the benefit of your dog?

During recreational walking, dog walkers are brought into connection with their surroundings and can appreciate nature, for example:

"I like looking at maybe a shaft of sunlight that comes
through the trees and lights up maybe one specific area
… or the change in the scenes."

In order to leverage the full benefits of dog walking, do it
mindfully. Don't just stick your headphones in or spend your
dog walks checking social media. Use your walking time to
be present, appreciate your surroundings and observe your
dog. Look at the leaves, listen to the birdsong, take some
deep breaths. By paying attention to the walk and what your
dog is doing, you are also more likely to be a responsible dog
owner – or *response-able*.[20] You are able to notice when your
pet is about to run over to another dog – just because yours
is friendly, doesn't mean theirs is – which may be frightened
by the approach. You are better able to react and control
your dog so that it doesn't negatively impact others and get
a bad name.

Having said that, sometimes we need extra help and motivation to get out on a walk, and this is where "temptation
bundling" can be helpful. I enjoy listening to podcasts, but I
really don't enjoy folding and putting away the laundry, which
would sit in piles on our spare bed for weeks if I let it. One
strategy I use to make me tolerate, and even look forward
to, doing household chores, is to allow myself to listen to
podcasts during these times. I've also learned to use this strategy when going for a run on my own or fitting in an extra,
early-morning dog walk, by downloading an audiobook that
I only get to listen to when I am performing these activities.
It is important that you don't spend all dog walks ignoring
your dog, for the reasons described above, but temptation
bundling with a more pleasurable distraction can be a useful

strategy when you need some extra motivation. For me and my dogs, that sometimes means we have an on-lead stroll first thing while I listen to an audiobook, and another more interactive off-lead run later in the day.

After reading this chapter, take a few moments to self-reflect and complete the table below, in particular the part about planning for "What if?" situations (if X happens, what will I do to get round it?):

PLANNING IN ADVANCE FOR CHALLENGES AND HOW I WILL OVERCOME BARRIERS TO DOG WALKING

My main challenges and barriers to more walking	Strategies I will use to address those challenges or barriers

CHAPTER 10

MORE TRAINING FOR AN EASIER LIFE

INTRODUCING THE DOGGY DICTIONARY

In this chapter, I will cover how to train some of the other commands (or "cues" in training speak) that can be useful for your dog to know. These can make your life easier, but it is up to you which commands you think will be helpful for your particular situation. You also need to decide which words mean what actions. You can train your dog to sit to "apples" and lie down to "pears" if you want. The top priority is that you know what the word is, and that everyone in the household uses the same word to mean the same thing. For this reason, it is helpful to print out a "doggy dictionary" and put it in a prominent place, such as the fridge door, for everyone to take note. There is also a column for describing a hand signal. We often forget that dogs communicate more visually than by sound. Your dog will find it much easier to understand what you want from him if you also give a clear hand signal.

DOGGY DICTIONARY

Word	Hand signal, if necessary	Response

Place this list where all the family can see it. Be consistent – it is essential that everyone uses the same words. Add to this list as your dog's vocabulary increases.

SIT

Remember to keep training sessions short and sweet. Each training session should be only a couple of minutes before you let your dog have a break. Like small children, dogs cannot concentrate for long. It is far better to end before they get bored but have learned something constructive, however small. How fast your dog progresses through the stages outlined will depend on your dog, but expect to have to do a few short sessions at each stage (over a few days) before you move on to the next. In the early stages, ensure that no one tells your dog to sit without having a big reward to give him when he does what he has been asked.

STAGE 1

Start in a quiet place without distractions, such as your living room, with just you and your dog. Decide which word you will be using and write this in the "doggy dictionary". By using the treat as a lure, position your dog. For a dog to sit, his back end must go on to the floor. Start with the treat in your hand placed near the end of his nose; he may try to nibble at it. By slowly raising the treat above his head and slightly backwards (aim just above and between his eyes), his head follows the treat and his back end should lower automatically (overleaf) (although this may take a few seconds). Once his bottom hits the floor, the treat can then be given as a reward.

STAGE 1: LURING THE DOG INTO A SITTING POSITION WITH A TREAT

If your hand is too high, he will jump up. If your hand is too low, he will stand still or walk backwards. Don't use your other hand to force him into the position. We are teaching him to want to do the right thing for us. If it seems to take a while at first, don't worry. Be patient.

STAGE 2

Note that until now, your dog has not been told the word "sit". Once you are 99 per cent sure that when you move your hand up and back, your dog will sit, start to introduce the cue word.

STAGE 3

Your dog should now be able to follow the hand signal without you actually holding the treat in front of him. (I suggest holding your palm horizontally with the back of the hand facing down towards the dog, then moving your hand towards you from horizontal to vertical. This mimics what your hand was doing with the treat when you lured him, but is sufficiently different from the "stay" signal we will use later.)

STAGE 4

By now, your dog should be responding to one command. If you tell him to "sit" more than once, you are teaching him to ignore you, and he is learning that "sit" means "do anything you like". Give your dog one word. If he then has not done what you wanted, don't simply repeat yourself over and over again. Stop and ask yourself why:

» Is it because he wasn't listening? If so, get his attention and try again. Work on getting his attention in different situations.

» Is it because he really doesn't know what you are talking about? If this is the case, go back a few stages.

» Is it because he has learnt to ignore you? If so, go back to having a treat in your hand each time you ask him to do something. Only ask something of him when you have his attention.

» Perhaps you are not consistent with your words, and he is confused because you sometimes say "sit" and sometimes "sit down" or even "come on now, sit."

» Is he perhaps in pain? If a dog is having any discomfort in his back legs, he will be reluctant to sit because it may hurt him.

» Is it because he gets more attention for not sitting than he does for doing as he is told?

» Have you taught him how to work for fewer treats? (Re-read Chapter 7)

» Is it because he doesn't want to do it? If not, why not?

» Is it for some other reason?

Only you can find out the reason why your dog has not responded in the way you wanted.

Each time your dog "misbehaves", ask yourself "why?" Once you know this, you will know how to put it right. Work on resolving the difficulty and do not simply be cross with your dog. He has good reasons for acting as he did – we need to understand these and then work on resolving the cause. Remember, we call it "misbehaving" or "being naughty", but from your dog's point of view, he is either behaving naturally or he is behaving how *you* have taught him to.

DOWN

When your dog knows how to respond to "sit", start teaching him to lie down on command. Be careful with the word you choose. If people are going to tell him "down" when he jumps up, or when they want him to get off the sofa, then perhaps use the word "lie". The alternative word for those situations is "off", and then you are free to use "down" to mean "lie down on the floor". This works well for me, but I still on occasion hear my husband telling our dogs to "get down" – and he wonders why they stare at him in confusion.

STAGE 1

Use a treat to teach him to lie down. Have your dog in a sitting position, reward him for sitting, and then produce another treat. Let him sniff this, but not eat it, and then slowly move your hand down to the floor so that you are holding the treat

STAGE 1: LURING YOUR DOG INTO THE DOWN POSITION WITH
A TREAT

between his front paws. Hold your hand still there and see if
he lies down to get at it.

Many dogs will lie down, but you may find you get better
results by moving your hand slowly forwards (away from
him) or slightly backwards (towards his chest), as some dogs
seem to rock back into a down position and others slide
forward. If you move your hand too quickly, your dog will
stand up. If this happens, simply get him to sit and try again.

As soon as he lies down, say the word and reward him
with the treat and praise. Resist the temptation to push your
dog down, or pull his front legs out. At best, dogs simply
learn to allow you to do this. At worst, dogs learn to pull
away from you.

STAGE 2

If you are getting nowhere fast with Stage 1, there is an alternative. It may be easier to hold the treat under a chair or low
table (for big dogs), or even sit on the floor and lure them
under your knees (with little dogs). This way, your dog has

to follow the treat, and has to lie down to reach it. As soon as he lies down, say the word and reward him with the treat and praise. Don't push down on them at all with your legs – simply let them crawl under you to get the treat they desperately want to get to. Once they have done it this way a few times, go back to the other method and they should pick it up.

STAGE 3 onwards

Similar to the sit command, reduce the luring with food into a hand signal. For "down", I suggest your outstretched palm facing the dog and then moving down.

STAY

In Chapter 8, we taught the dog to "wait", which is different from "stay". Teach him that "stay" means "do not move from that position and I will give you wonderful treats". It can be useful as another foundation exercise to teach your dog, to help you learn how to train, and is often necessary for any Canine Good Citizen or obedience testing.

Being on that chosen spot results in treats, praise and fuss. Being off the spot results in those good things ending, and simply being returned to the spot.

STAGE 1

Begin with your dog in a sit. Ask him to "stay" (hand signal is an outstretched palm, like you are asking him to "stop") and then after one second, give him the treat. Then give your release command word (same as you chose for "wait") and let him get up and move around.

STAGE 2

Gradually increase the time your dog will remain in the sit position. Begin with one second, then a couple of seconds, and gradually increase a few seconds at a time. If he gets up, do not reward, just lure him back to exactly the same spot he started on and ask him to sit and stay again. Progress to ensure that he is happy to stay near you, *without you moving*, for a whole minute.

Teach for a short period of time, always praising him in the position, *before he moves*. If you praise him after he gets up, you are teaching him not to stay, and he is then getting the reward for moving or for coming to you, which we don't want.

STAGE 3

Only now should you start to progress to moving slightly away from him. Take one step away, then step immediately back to his side, praise, reward and release.

STAGE 4

Once your dog is responding well to this, and happily not moving, try taking two steps away and then back.

STAGE 5

Once your dog is happy to stay calmly while you move a few steps away, and separately can stay for a good period of time with you next to him, begin combining time and distance. However, don't expect your dog to be able to do both straight away to the same level. Reduce time and distance to begin with, starting from well below what you might expect your dog to be able to do separately.

STAGE 6

When he is happy to "stay" in the sitting position, teach him to stay while he is lying down.

FURTHER STAGES

Can your dog stay without moving while you walk behind him? While you bounce a tennis ball in front of him? While you leave the room completely? While you lay down on the floor and then get up again?

HAND SIGNAL OR VOICE?

You think you have taught your dog to sit and lie down on command. Give your dog one word, such as sit or down. Notice what he does. If he has not responded, find out why and resolve the difficulty. Do not simply repeat the word over and over again or shout louder. This will only teach your dog to ignore you. One reason why could be that he doesn't actually know the word as well as the hand signal you were cueing him with during training. Now try giving the same command, but only using the hand signal, and no voice command. Did he respond this time? If your dog is responding more reliably to the hand signal than the voice command, I would say you are in agreement with 75 per cent of the dogs and owners in my classes. In order to improve, practise this exercise three times in a row: 1) Voice and hand signal (and reward); 2) Voice and hand signal (and reward); 3) Voice only (and hopefully big reward and fuss). Because your dog has just been successful at doing the same task, he should have a good idea what you want by the time you ask the third time – without the clue of the hand signal. If he hasn't got it yet, just keep practising in short sessions until he does.

NOT JUMPING UP

A dog will do what is most rewarding, so if your dog gets more attention for jumping up, he will jump up even more. Make sure that you give praise and fuss him only when he has all four feet on the floor; it is essential that visitors do the same. Teach him to sit for attention, and ignore him for jumping up. Use every available opportunity to teach him. He will be persistent; you must be more so. If someone wants to greet your dog and says they "don't mind" if he jumps up, tell them you do! Remember the power of intermittent rewards – it's better for motivating unwanted behaviour than if the dog was rewarded every time. If you know someone is going to break the rules, then have your dog on a short lead and step on it to hold the lead down to the floor with just enough length that the dog can comfortably stand but will be prevented from jumping up. The dog isn't really learning anything, but it's better than having your good learning undone.

HANDLING

The only time many dogs are intimately touched is when there is something wrong and they are in pain, when they are being brushed or bathed, or when they have to go to the groomers with a knotty coat. Clearly, these are not the best situations for your dog to learn to tolerate being handled, so you need to begin work on this much earlier.

STAGE 1

Teach your dog to be handled all over by you. Begin by using your hand and ensure that you can touch him all over without him resisting, mouthing, getting frightened or turning it into a game. If your dog does any of these things, have a treat in your other hand (or another person with a treat) and aim to do a little at a time. Instead of trying to handle him everywhere, simply stroke down his paw and then give him the reward and leave it at that. A little while later, show him another treat and look in one of his ears, handle it briefly, and then reward him.

The secret to having a dog who is calm to be handled is to reward the good behaviour – the staying calm and allowing it to happen. The reward is a treat, but the main reward is you stop handling them (as that's their ultimate goal). Therefore, do not allow your dog to squirm, mouth or resist. By "do not allow", I mean don't let it get to the stage where they do any of these things, because you have already stopped and rewarded them before they resisted what you were doing. This is why the actual handling is incredibly short to begin with – it is a training exercise; you are not actually trying to inspect their paw or eye at this stage. As they become more patient, and happily wait for the treat, you can extend the duration you are touching them.

STAGE 2

What part of your dog is the hardest for you to handle? Spend a little more time teaching him that he likes to be touched in this area. Spend lots of short periods of time teaching him how you want him to react, and reward the good behaviour.

When Roxie was a puppy, we were doing these same exercises in Erica's class. Roxie was tired during class one day and quickly learnt to growl at me to stop me looking in her mouth. We were both a bit shocked by the situation, and I had to spend the next few weeks teaching her that it was rewarding for her if I looked in her mouth. To do this, I briefly touched her chin, let go and then (before she could growl) gave her a treat. Over the next few weeks, we slowly built up the amount of time I handled her mouth, until she would let me pull her mouth open and take a good look inside. She still doesn't exactly like anyone looking in her tiny mouth, but she lets us anyway as she knows it won't be for long and she will be rewarded afterwards.

STAGE 3

Your dog should now be happy to let you handle him all over. When you can do this with your hand, start to teach him how to behave when you touch him with a brush. Again, take your time and progress slowly.

STAGE 4

Have you ever noticed that when a dog gets a cut or a lump, it is nearly always on his stomach or somewhere difficult for you to examine? Teach him to accept being rolled over when you handle him. Again, use treats at first and gradually teach him to be happy in this position while you fuss him. Keep practising.

STAGE 5

Now concentrate on teaching your dog how to behave when dried with a towel. The worst time to teach this is when your dog is wet or muddy. Teach him when he does not need to be dried and then you can "dry" one leg, reward him and leave it for that session.

STAGE 6

Your dog must also learn to accept other people handling him, such as the vet, groomer or show judge. Use praise, treats and games to reward the correct behaviour and to make it enjoyable for him. You cannot teach him this alone, so get other people to help. Start by using people whom your dog knows and likes, and then progress to those he does not know.

STAGE 7

People tend to grab a dog by his collar. This is not always a sensible thing to do, as many dogs object to this. However, it may happen to your dog, either with you, or your family or a stranger. It makes far more sense to teach him now that being grabbed by the collar is pleasant. To do this, touch his collar and reward him. Then grab his collar and give him a treat, grab him and gently pull him towards you, give him a treat then and tell him he is a good dog, etc. When you call him to you, touch his collar before you reward him. Obviously, do not grab so quickly that you frighten your dog, but teach him to accept that this is pleasant. If possible, get other people to touch his collar and then reward him.

STAGE 8

Ensure that you can open his mouth, not just in the house but also when you are outside. Ensure that he is happy for you to look in his paws and down his ears when he is standing and when rolling over. It is often easier to handle your dog when you are inside the house, but think about when you are on walks – your dog may pick up something you have to remove from his mouth, he may cut his paw, hurt his leg, etc. Therefore, it is essential that he is used to you stopping on walks to handle him at times.

SEND TO BED

Sometimes it is useful to be able to send your dog to his bed. This should not be used as a punishment, but as a rewarding place to be that is out of your way while you are doing something. For this to happen, it needs training.

STAGE 1
Hold your dog by his collar and show him a treat that you are placing on his bed. Then encourage him to come a few steps away with you, point to the bed and say "bed" and let him go – he should run to the bed and be rewarded with the treats. Increase the distance your dog is running to the bed over time.

STAGE 2
Now that your dog has the idea that being on the bed means treats, pick the bed up and with him watching, put it on the floor in front of you and him. He should run over to the bed, at which point you say "good boy" (or use your clicker) and toss him a treat in the vicinity of the bed. Practise many times, moving it around the room, until your dog is reliably jumping on the bed in expectation of a treat.

STAGE 3
This time, leave the bed where it is and toss the treat away from the bed and the dog. Your dog should move away, eat the treat, and now ... you wait. By this stage, your dog should be clearly making the connection, and moving onto the bed of

his own accord, looking at you in expectation. Practise until your dog is reliably running to the bed of his own accord.

STAGE 4

This time, don't immediately reward your dog. Wait a few seconds and see what he does. Hopefully, in his frustration that the treat isn't appearing, he should sit or lie down on the bed to see if that works, at which point you can praise and reward. If this isn't working, spend a few moments luring the dog into a down position on the bed. From now on, he only gets the treats if he lies down on the bed when he gets there.

STAGE 5

Now that the dog is reliably running to the bed and lying down, give him the command "bed" before you send him (pointing to it also helps).

STAGE 6

Now that your dog reliably runs to the bed, you want to teach him to stay there. Now, rather than toss treats at the bed, walk and reward him there, without him getting up. Teach him that if he waits in the down position longer, more treats come after the first one.

CHAPTER 11

WAKE ME UP BEFORE YOU GO

SPECIFIC TRAINING TASKS FOR HEALTH

Whether or not you have specific health requirements, there are three tasks you can train your dog to perform that I would recommend to anyone. Firstly, train your dog to wake you up, like a canine alarm clock. Secondly, train him to find and retrieve specific items on command – your phone or medication pouch, for example. And thirdly, best of all, train your dog to cuddle on command. All of these can be achieved through reward-based training, and ensuring that the dog finds helping their owner fun to do. Although I will be describing in some detail how to train tasks that can help with your health, please remember to adapt the methods somewhat to your own needs – such as the specific positions you need to be in due to your physical abilities and constraints. I would also advise that you don't start any health-task training until your dog is at a sufficient age and maturity. These are fairly complex tasks to learn (from a dog's point of view – compared to earlier tasks like sitting or lying down on command) We also would not want to put undue pressure on growing joints through actions such as jumping on and off the bed or holding an alert position. Let your pup be a pup, and wait until they are at least eight months, or even a year, old (and therefore hopefully past adolescence and any sensitive periods).

BACKCHAINING

When I was fresh out of university, I was interviewed for a job training dogs to assist people with a physical disability. On the day of the interview, I had to travel many hours to the training site, and had a stinking cold. It's fair to say I wasn't at my best, which may partly excuse why, when asked if I had any questions for them, I asked "Just how *do* you train a dog to load the washing machine?" Needless to say, I didn't get the job. In fact, I didn't even hear back from them to say I *hadn't* got the job, it was that bad. I've come a long way in my learning since then (and did soon after get a job with a different organization). I know now that whenever a behaviour looks like a complex task for dogs to be trained to perform, it is actually a series of much smaller tasks, put together backwards. This principle is called "backchaining" (opposite).

FIND A HELPER

In order to train most tasks, you will need a helper. A dog can be taught many things by just one person, but an extra person will save you a lot of time and heartache. Jake was

BACKCHAINING

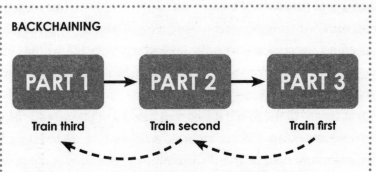

Essentially, the end of the sequence is trained first. Once the dog is confident at performing that small end part for a reward, a new and immediately earlier part of the sequence is separately trained, again for a reward. Once the dog can perform the new part well, the two tasks are put together. Because the dog already knows the end behaviour well, it can be cued to do it immediately after the earlier part ends, and run straight through to get the reward it knows is coming – and so on and so on. In this way, a series of complex tasks can be chained together for just one end reward.[1]

a terrier cross that I was training. He was bright and motivated, but late on in his training I was getting frustrated; I felt he knew what he needed to do but was reluctant to do it. When he heard the cooker timer that he was supposed to tell me about, he would look at me, but then go off and sniff. I could encourage him to come to me, but he didn't seem to have the confidence to do it of his own accord, which is essential because his deaf owner would not be able to hear it. For some reason, Jake lacked confidence, and my frustration at him was making things worse. I realized that during his weeks of training with me there had been a difference from my usual methods: we much less often had someone to help us with our training, and mostly trained alone. Without this person's assistance – to hold Jake away from me so that he was motivated to run towards me, and then to encourage him back to the sound he was taking me to – Jake was less confident to run to me to tell me he had heard something and then would appear to "forget" where he was going when I asked him what he had heard. Thankfully, the problem was easily fixed with more training (with a helper) and I made sure I didn't make that mistake again.

> **NOTE for all training**
>
> Only ever practise for short periods of time – five or six minutes at the most per session. Your dog needs to enjoy the sessions and be able to concentrate. Remember to be sure that your dog is really confident about the present stage before you make it any harder. If your dog struggles at any point or seems confused, go back a stage in your training until your dog is confident again at that level, and then move on to the harder level.

ALARM CLOCK

It can be hard dragging yourself out of bed in the morning, especially if you are feeling depressed. If you have hearing loss, you might not even be able to hear the high-pitched beeping of an alarm. Dogs are usually enthusiastic to start the day, and need to go outside for a pee and then want breakfast. So this task is an excellent one to train, from the dog's point of view. You will need to be happy with your dog sleeping in the bedroom, though.

Keep a pot of treats (or a toy) by your bed where your dog cannot reach them, so that you can reward your dog when they do this task well. Decide whether you would like your dog to jump on you in bed and scrabble at you when the alarm goes off, or whether (if you have a large dog) you may prefer him to stand at the side of the bed and put his front paws up and nudge you with his nose (may be less shocking than a large dog landing on you while you are in a deep sleep). How you train this task will be different depending on the method you choose. We will train the end of the sequence first, which would be the way the dog gets your attention when you are asleep.

Jumping on the bed and scrabbling
STAGE 1
Encouraging your dog to jump on the bed is important for this task. Start by having your dog held gently by the collar by someone they know and trust, in the place they would normally be sleeping (e.g. on their bed in the corner of the room) while you are lying on top of the bed, on top of the covers. Show

A SMALL DOG JUMPING ON THE BED AND SCRABBLING TO WAKE THE OWNER WHEN THE ALARM CLOCK SOUNDS

your dog that you have a treat, and call them to you. Ask the person holding your dog to let them go when your dog is keen to come towards you. Keep encouraging your dog to jump on to the bed as they run towards you. If they are unsure, encourage them by showing them the treat and moving it towards you on the bed. When your dog jumps on to the bed, reward them.

Sometimes give your dog one treat and other times wait a few seconds and give them a second (or even a third) treat. Practise this until your dog is confidently jumping on your bed, and make sure that they will do this from various places around the room.

STAGE 2

Begin to do this again while you are under the covers. When the dog jumps on the bed, encourage them to jump onto/next to you. Once your dog is doing this well, start encouraging them to scrabble their paws on you, by withholding the treat that they are expecting. They should get a bit frustrated that you aren't providing the treats as before and work harder for it – but try not to encourage them to scrabble at your face. When your dog is jumping up confidently on to the bed and scrabbling at your body parts, move to Stage 3.

STAGE 3

Now start to add the alarm clock sound. Make sure that the dog has heard the sound before and is not worried by it. When your dog jumps or puts the paws on to the bed, turn on the alarm clock (that you are holding at this point) so that they learn to associate the sound of the alarm with being fed treats on the bed. Do not worry if your dog is a bit distracted by the sound at first, this is normal.

STAGE 4

When your dog is confidently jumping up to the bed and scrabbling/nudging with the alarm clock sound going off, start to introduce the alarm clock sound just as your dog is let go to run towards you.

STAGE 5

Set the alarm clock noise off just before the dog is let go by the helper. Practise this from a variety of places in the room and until your dog confidently runs up towards the bed, jumps up and scrabbles/nudges at you until you turn the alarm clock off, after which you give them a food reward and get out of bed. Remember to vary the amount of treats you give to your dog and also the length of time that they scrabble or nudge before you finish the exercise each time.

STAGE 6

You can then begin to practise this "for real". With your dog waiting on command in his bed this time (which you should practise beforehand, using the guidance in Chapter 8), set your alarm for 10 seconds, pretend to sleep deeply and see if he reacts when the alarm goes off. If not, go back and practise the earlier stages again. If he does well, work on extending the fake-sleeping waiting time. If he breaks the wait too early, don't tell him off (and risk discouraging him from reacting to the alarm clock) – just calmly put him back in the wait position and choose a shorter time for now.

STAGE 7

Set your alarm clock for the morning. When the alarm clock goes off, hopefully your dog will respond. If not, give a little help and encourage your dog to jump up. If the dog is unsure, do an easy, short practice exercise again immediately, so that the dog finishes feeling confident about the task.

Nudging with front paws on the bed
STAGE 1
Encouraging your dog to put their front paws on to the bed is important for this task. Start by having your dog held gently by the collar by someone they know and trust, in the place they would normally be sleeping (e.g. on their bed in the corner of the room) while you are lying on top of the bed, on top of the covers. Show your dog that you have a treat, and call them to you. Ask the person holding your dog to let them go when your dog is keen to come towards you. Keep encouraging your dog to run towards you. If they are unsure, encourage them by showing them the treat and moving it towards you on the bed. Use the treat to get your dog to put their front paws on the bed. When your dog puts their front paws on to the bed, reward them.

If they are too excited and jump full on the bed, don't tell them off. Just guide them on the next time and call them with a little less excitement.

Sometimes give your dog one treat and other times wait a few seconds and give them a second (or even a third) treat. Practise this until your dog is confidently putting their paws up on your bed, and make sure that they will do this from various places around the room.

STAGE 2
Begin to do this again while you are under the covers. When the dog puts their front paws on the bed confidently, start

encouraging them to nudge you with their nose or paws. If you withhold the treat for a few seconds, they should get frustrated and begin to nudge you with their nose or paws for attention, which you can then reward. When your dog is reaching up confidently on to the bed and nudging at you, move to Stage 3.

(Complete Stages 3 to 7 starting on page 222)

PROBLEM SOLVING

Sometimes a dog who has been previously well trained to perform an alarm clock stops doing it. Usually, this coincides with them being given a really comfy dog bed to sleep on, and they can't be bothered to get up. The solution may be to downgrade the bed a bit, to one that is comfy but not as luxurious. More importantly, you should increase the value of the reward he gets for leaving his bed. You may also wish your dog to actually sleep on the bed with you. Be aware that the same issue may apply to this – that the dog may be too comfy to move – but depending on the dog (and the reward they get), this can work well. Train exactly the same way and hopefully the dog will still climb onto you and nudge/scrabble to wake you when he hears the alarm.

FURTHER DEVELOPMENT

Perhaps you want to go even further and train your dog to pull the bed covers off and pull the curtains open if you don't respond.

PICK UP/RETRIEVE

Picking up items that we have dropped or cannot reach can be helpful for those with health issues. To teach this, you will need a dog toy that is safe for your dog to pick up, and later a soft item like a purse, to act as the "object" for this task. (A suitable soft toy on a keyring can be a useful starting point to teaching your dog to pick up a bunch of keys.) Your dog should not automatically pick up the item you have dropped (just in case it is sharp or otherwise unsafe for your dog), but only on cue. Some dogs will retrieve items easily with little training, while others will need a lot of help as it will not come naturally to them.

STAGE 1

Get your dog excited about the toy (and hopefully interested in trying to grab it with their mouth) while you have it in your hand. Have a supply of favourite treats in your other hand; they must be something that your dog likes more than the toy. Then, drop the toy on the floor. If your dog automatically picks up the toy straight away, praise (with a click or "good boy", or whatever your marker word is) and give them a food reward. If your dog does not automatically pick up the toy, point at the item and try praising and rewarding them when they simply touch the object, rather than having to pick it up yet. Keep doing this until your dog is actively trying to touch or mouth the object. Practise this and keep encouraging your dog until they try to pick it up, and once they are confident at this, change to only rewarding actually picking it up. Once they are picking it up confidently, you can start saying "pick it up" as they do it, to make the association of the cue with the action. If your dog is not super confident at holding on to the toy at this stage, try not to take the object away from them straight away. You want to be rewarding the dog for holding the item as long as possible, even when you are reaching out to them, so pinpoint this with your marker word or clicker.

STAGE 2

Teaching a dog to "give" the toy back on command is vital. Choose a word to use, such as "give", "thank you" or "drop it". Teach him that "give" means "I have something really nice in my hand, and if you let me take that toy, I might give it to you". This means that you are not forcing him, but simply rewarding the good behaviour.

If you have any difficulty in getting the dog to bring the toy back to you and/or give it to you:

1. Have a line or lead on your dog when you are playing, so that they can't run off with the toy.
2. Have two toys, so that whatever he has, you have something better. These may need to be identical toys, so that your dog is less likely to decide that he prefers one.
3. Practise in a small area so that they can't run off with the item.
4. Train in the corner of the room, with the item in the corner, so that they have to pass you with the item.

Sometimes, being possessive with toys can be a symptom of other difficulties. If your dog really does not want to give the toy up, contact a trainer for advice.

STAGE 3

Each time they take the toy into their mouth, tell your dog to "pick it up". Just before they drop it, tell them to "drop it" or "give" (whatever you taught earlier), but now put out your hand for them to drop it into, and reward them. Practise this so that you wait longer each time that your dog has the object in their mouth before rewarding them, but if your dog begins to drop the object before you reward them, take it back a stage. Do not worry at this stage if the object falls out of your hand or on to the floor after your dog has dropped it near your hand. Attempting to get it in your hand is what we are currently rewarding.

STAGE 4

When your dog is confident in picking up and holding the toy, and trying to drop it in your hand, now only reward your dog for the "drop it" when your dog drops it into your hand. To begin with, it is probably easier for the dog to do this when you are sitting, but remember to practise standing up as well. From now on, make sure that you always practise the "drop it"command into your hand, so that your dog knows that the toy must be put into your hand. If it is not, they have to pick it up again before they will get the food reward.

STAGE 5

Now, make the task harder by dropping different items that have different shapes and textures, such as a small bunch of keys, a small box, or a tissue. Because these items are harder for your dog to pick up, you will likely have to go back to earlier stages of training with these items instead of the toy, and work your way back up through the stages again. This is normal and to be expected.

STAGE 6

Now try dropping a simple, easy item, such as a soft toy – but as you are moving along, for example walking, and ask the dog to pick it up and follow you to bring it to you. Then progress to more difficult items. If your dog begins to struggle, make the exercise easier by not moving so much or by using an easier item.

STAGE 7

Can you train your dog to recognize the names of different items and bring them to you, rather than just retrieving whatever you are pointing at?

CUDDLE/TOUCH

Many owners love to receive tactile comfort from their pets, and will stroke them and hug them, especially when they are feeling sad, in order to feel that rush of oxytocin. But have you ever asked your dog if he likes to hug? To test it out, try this task. (NOTE: If you already know your dog doesn't like being hugged and cuddled, obviously don't do this.)

1. Sit on the floor with your dog. Cuddle your dog for a few seconds, and then stop. How did your dog react? When you were hugging your dog, what was his body language – was he stiff? Could you see the whites of his eyes? Were the ears flattened back? Did he turn his head away from your face? If so, perhaps your dog doesn't like cuddles. Alternatively, did he push his head into you? Was his face and body language soft and floppy? When you stopped hugging him, did he nudge and lean into you, or paw at

you, to try to get you to cuddle him again? If he did, your dog probably likes being cuddled.

2. This time, don't lean into your dog like you would for a cuddle, but outstretch your hand away from you and give them a scratch instead, perhaps under their chin or belly. What is their body language like now? Did they enjoy this more than the hug? After a few seconds, stop the interaction and wait. Does your dog indicate that he wants you to stroke him again? If so, perhaps they enjoy less intense physical contact with you than hugging.

If you conclude that your dog doesn't like cuddles, do not despair. There is an alternative interaction you can teach your dog – to touch your hand with his nose for a reward. This is also useful to teach any dog, whether they like hugs or not.

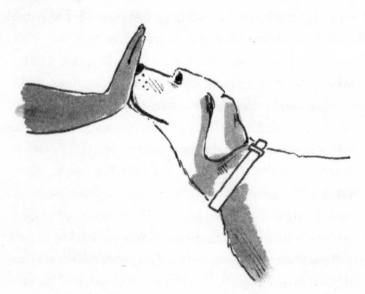

NOSE TOUCH TO HAND

STAGE 1

Hold a treat in the palm of your hand under your thumb. Let your dog sniff it out and as they nudge your palm, praise them (with a click or a "good boy", or whatever your marker word is) and let them eat the treat. Repeat until your dog is doing this confidently.

STAGE 2

Offer your outstretched hand in various positions and heights, so that the dog is coming over to nudge your palm in many different locations of the hand (high, low, left, right), and from different distances.

STAGE 3

Offer your hand without a treat in it. If the dog nudges your palm, praise and quickly give the reward from your pocket or treat bag. If not, go back and practise Stages 1 and 2 until your dog is ready for this stage.

STAGE 4

Once your dog is confidently nudging your palm in many varied circumstances, start to give the command "touch" or "hand" just before, or as, he does it.

STAGE 5

Practise holding out your hand and telling your dog to "touch" in a variety of situations – indoors, outdoors, on a walk. This is a great one to get your dog to focus on you in a distracting environment.

STAGE 6

Once your dog is adept at this task, withhold the treats for a few seconds so that your dog becomes even more persistent, keeping his nose in your hand for longer, or even nudging multiple times.

STAGE 7

Whenever you feel you need some comfort from your dog, you now have a command to get them to come over and physically touch you.

If your dog likes cuddles, then it is possible to use the method above to also train the dog to "cuddle" on command. The same principles apply, you just reward a different (and more enthusiastic) behaviour. If your dog is large, maybe you wish for him to just put his front paws up on you and snuggle across your chest. If your dog is small, perhaps you train her to jump up on your lap and put her paws on your chest and nudge your face. To start with, reward anything resembling the early stages of the behaviour you want, and then ask more of the dog as they get more confident (by withholding the treats a few seconds until they get frustrated and offer more behaviour). Once you are sure that your dog is repeating the behaviour well, introduce the command that you want to use and associate it with the behaviour.

TRAINING TIPS

» Training with your dog should be fun – you are a partnership and it is important that you both enjoy working together. If you are struggling or feel disheartened, contact a trainer for advice.

» If, at any time when you are training an exercise or a task, your dog does something you don't want them to do, never tell them off. Just ignore the behaviour you don't want and encourage them to do what you are showing them.

» Always finish any exercise at a point where the dog has done something well and you can reward them for it. It is important that you and your dog come away from the training session happy and relaxed, and that your dog is excited to do it again next time.

» If your dog is struggling with an exercise, practise something easier, reward them for getting it right and then have a break.

» Try not to get frustrated with your dog if they are struggling – have a break from training and try again later, or contact a trainer for advice.

» Only ever do short training sessions so that your dog does not get too tired. At first, any training exercise will be tiring for your dog, as they are not used to doing these things and they will have to learn a lot. Make sure to take breaks during longer training sessions and have a game or fuss your dog.

» Remember that it is important to treat your dog sometimes for doing tasks even when they are experienced – this keeps them motivated and excited about any task they do.

» Always vary the number of treats you give your dog during training for different tasks and exercises. Give the treats individually. This encourages the dog to keep doing the task, as they won't know how many treats they will get.

» Once your dog is trained and is performing a task well, your job isn't over. You will need to regularly practise using training sessions to keep your dog working effectively in real-life situations.

CHAPTER 12

FURTHER ASSISTANCE TRAINING

There are many other helpful practical tasks that a dog can be trained to perform: reminding you to take your medication, closing and opening doors, interrupting unwanted behaviours, or letting you know someone is at the front door. The choice is yours, and the only limit is your imagination, but the reward-based training principles remain the same. We will discuss potential approaches to some of these tasks later in this chapter.

THE DIFFERENCE BETWEEN ASSISTANCE DOGS, EMOTIONAL SUPPORT DOGS, THERAPY DOGS AND PET DOGS

If you have a physical or mental health need for which you can imagine that having a dog trained to perform a number of different tasks would be helpful, it may mean that you would benefit from having a full assistance or service dog (the latter being the term often used outside the UK). Through the rights of a disabled person in most countries to have reasonable adjust-ments made for their disability, recognized assistance dogs are allowed by law in public places that pet dogs are not, such as shops, restaurants and taxis. Assistance dogs are typically bred and trained by an experienced charity or other organization, before being matched to and placed with a suitable handler, and are provided ongoing training and support by the organi-zation. This is done in order to safeguard the welfare for both the dog and the handler by ensuring that the organization uses respected (and hopefully kind) training methods, and that the dog has the correct temperament and behavioural standards to do the challenging work.

"Emotional support dogs", on the other hand, have typi-cally had no specific training for the role, and are not usually accepted to enter public places like assistance dogs are – which is a key distinction. However, in some countries they are "recognized" in that a person can designate their pet dog an "emotional support dog" for the purposes of being allowed pets in a rental home or to take in the cabin on a plane. They often do not meet the behavioural and temperament levels that would be expected of an assistance dog in a public place (and anecdotally, at least, some are often nervous or even aggressive and a threat to the public). Their owners may argue that the animal's presence makes the owner feel better, so they need their pet with them, but many pet owners would say exactly the same. I would turn that question round – does being taken by their owners into often stressful situations, such as on planes or in shops, without specific breeding and training for this role, make the dogs feel good? Is this fair?

A therapy dog is another different thing entirely; they are assisted by a handler (usually their owner), rather than assisting the handler with their needs. The handler and dog should receive training and qualification by a recognized provider and then may go (upon invitation only) into places such as schools, hospitals or care homes to visit the patients or children there, and provide opportunities for interaction with what is essentially a pet dog with a really nice temperament. In fact, the scientific and practising community doesn't really like the term "therapy dog" either, as it would be a more appropriate term for a dog who works alongside a professional counsellor or therapist, which most do not. It also oversimplifies the complexity of how animals are used by professionals in this field. The work of most "therapy animals" is more accurately described as "animal-assisted activities" (AAA), "animal-assisted interaction" (AAI), or "animal-assisted therapy" (AAT) depending on the context, but I use the words "therapy dog" here, as that is what most of the general public would call them.

Like research into the effect of pets on their owners, surprisingly little research has been done into the effects of assistance/service dogs on their handlers. Qualitative case studies suggest that their handlers do believe that their assistance dogs improve their quality of life, and those handlers that are waiting for a dog also expect them to.[1] Recent research by Dr Maggie O'Haire, at Purdue University in Indiana, is trying to examine whether service dogs actually improve the mental health of war veterans with post-traumatic stress disorder (PTSD). Compared to those on a waiting list for a dog and receiving "usual care", handlers with a trained service dog showed clinically significant reductions in PTSD symptoms, including improved depression, quality of life and social functioning. However, despite these improvements they still remained in the diagnosis category for PTSD[2] (but then we don't expect guide dogs to restore the sight of their handlers either). Similarly encouraging positive findings have since been recorded for handlers with dogs to assist with physical or medical conditions too,[3] although a recent systematic review of studies of the impact of physical assistance dogs on their handler's mental health found that evidence of a positive effect was variable.[4] Given that owning and training an assistance dog is arguably even harder work than owning and training a pet dog, perhaps we shouldn't be too surprised by this.

Although in this book I am going to explain how to train a number of typical tasks that assistance dogs might be expected to do, there is a big difference between pet dogs and assistance dogs that needs some reflection. You should be well on the way to understanding now that it takes a lot of specialized breeding, socialization and training to produce a dog that can cope with the stressful and busy environments in which an assistance dog is expected to work. Most pet dogs wouldn't enjoy it. In fact, dog welfare has been identified as an important and often neglected aspect of research into assistance dogs.[5]

In the UK, it is typical for the handler of the assistance dog to not pay for the training, which is funded through charitable donations to the organization. This is an important reflection of the need to reduce conflicts of interest; this way, there is less personal and monetary investment if the dog being trained turns out to not be suitable for assistance-dog work and needs retiring from the programme. Although most dogs can be trained to perform simple tasks that are helpful to an owner, it does take a very particular temperament in the first place for a dog to become an assistance dog, which is expected to be "working" a considerable portion of the time. Obviously, the dog must have no health or medical issues themselves, as it would be unfair to make them struggle to work, especially when in pain. They must also have no behavioural problems – no signs of nervousness or aggression in any situation – but it goes beyond this. One example I wish to mention is Polly, a pretty little cross-breed who had passed her training period, but once she was placed with a handler, she kept having to come back for more instruction. Her demeanour was constantly sad. She had no problem doing the tasks asked of her to an excellent standard, but she was obviously not enjoying it. Many trainers worked on her, and new handlers were tried, but eventually the organization made the decision to retire her, despite the thousands of pounds that had been spent on her training. Once the pressure was off and she was no longer expected to "do" anything in particular other than be a dog, her demeanour changed completely. She was joyful and excitable, and lived a long and happy life with the family of one of her trainers. She simply didn't want to be an assistance dog, constantly expecting to have to respond at any moment. Even dogs need career counselling!

As Polly's story exemplifies, sometimes training a dog to assist our health doesn't go to plan, and we must remember to put the needs of the dog first. If you want to train your own dog as a full assistance dog, this is possible in some cases. Having the support of a skilled organization or charity is important to help navigate these sorts of conflicts of interest, including whether the dog is really suitable for the work, and when the dog deserves a well-earned retirement. Assistance-dog training and placement is complex and comes with many challenges that hinder or delay benefits to the handler. These include those arising from complexities of medical conditions, cognitive ability and skill of the handler, and the social environment the dog is expected to work in, as well as dog-related factors. [6] If it isn't done carefully and with support, the dog can bond to the wrong person, get confused, and be unreliable at performing the tasks it is meant to do – which is a risk to the very person it is supposed to be helping. It is for all these reasons that it is not recommended that you train an assistance dog by yourself. By all means, have a go at training your dog in a small number of tasks selected from those suggested here, but closely monitor their welfare and don't put pressure on them to have to do it well. If they are able, great. If not, they are still your wonderful pet dog. With all that said, let's look at what else we can train our dogs to do.

TRAINING AN ALERT BEHAVIOUR

An alert is a foundation for many training tasks. It means your dog has something to tell you, and will be rewarded for letting you know. An alert-type behaviour can be trained by either using a cue word or a specific sound. For the latter, you would need some object that is small, can be easily hidden in your hand, and makes a distinctive and reasonably loud noise easily, and when you want it to. This noise (or specific "alert cue" word) should only ever be used in your training at the appropriate time, or it loses its potency.

You also need to decide what kind of alert your dog will use (overleaf). Whatever method, it needs to be markedly divergent from your dog's usual behaviours. If you have a small dog, you could teach them to reach up and "scrabble" at you with both paws (and use the word "scrabble"). If you have a medium or large dog, you may wish for them to sit and place one paw deliberately against your leg, arm or wherever they can reach (and use the word "leg" – "paw" is probably too commonly said to your dog). If you have a very large dog, you can teach them to nudge you with their nose (and use the word "nudge"). If you choose the latter, the dog will need to continue nudging you with their nose so that it is very obvious that something is happening that you need to know about. The choice of alert will need to be suited to the physical characteristics, conformation and health of the dog, and also to that of the person (e.g. fragile skin, balance problems).

DIFFERENT TYPES OF "ALERTS" FOR YOUR DOG TO SIGNAL THAT HE HAS SOMETHING TO TELL YOU

STAGE 1

First check that your dog can hear the noise or sound, herein referred to as the "alert cue", and is not frightened by it. Perhaps alert cue and give them a few treats. Also try making the alert cue when you are not facing them, and check for a reaction.

STAGE 2

Sit down somewhere comfortable where the dog can easily get to your legs. Take out a treat and show it to your dog. Lure your dog into the chosen position using your hands and small treats. This will probably mean holding your handful of treats against your dog's nose and then moving your hand to your leg and exploring moving your hand around and wiggling it until your dog offers a behaviour that vaguely resembles what you are trying to achieve, in their attempts to try to get access to the treat. When they do, use your clicker or marker word to mark the behaviour and give your dog the treat. After a few times, tell them "off you go" and let them mooch around for a while and have a rest.

STAGE 3

Repeat the above as many times as necessary until your dog quickly touches you in this way as soon as you show him the treat. Sometimes give your dog only one treat, and sometimes give your dog two, three or four treats (one after the other), so that your dog begins to touch you for longer before going away (remember, the treats must be small; the size of half a little fingernail is more than enough). If you are teaching a scrabble and the dog starts to move their feet when they are touching you, quickly give them another treat so that they know that this is good. Repeatedly practise until your dog is reliably performing that behaviour to access the treat you have.

STAGE 4

Practise this with you sitting in several places. Once you are happy that your dog understands what to do, start teaching the dog how to do this when you are standing up.

STAGE 5

Once your dog is happy to consistently perform the alert behaviour, introduce the alert cue. When your dog is in position, give the cue. Your dog may be surprised at first and perhaps become distracted – don't worry about this, it is normal. Just carry on with the exercise and remember to reward your dog when they give the alert position. Keep doing this until your dog knows that the alert cue only happens when they are in the alert position. Build up the length of time that your dog alerts to the sound of the cue for, and add in the potential various places they will do it (rooms and positions), by continuing along the same process as above.

STAGE 6

Now, start to take away the food lure. At first, ask your helper to gently hold the dog's collar a little distance away from you, so that the dog is excited and wants to run to you when they hear the alert cue. When you give the cue, encourage your dog to run to you and alert. As soon as your dog starts to touch you in the correct position, repeat the cue and give your dog a treat from your pocket. Gradually practise this in more difficult scenarios, so that you can give the cue at any time, or anywhere in the house, and your dog will come to find you and alert.

When you have trained this exercise to a good standard, you may begin to use it as part of the tasks that you will be teaching your dog.

DOORBELL

I would be hard pressed to find a dog that doesn't react in some way to the sound of the doorbell or a door knock. However, typically their reaction is not that specific, but more crazed and based around running and barking. This may be enough for your needs to know someone is at the door (it certainly is in my house, where our "front" door is actually a back door hidden in a place the dogs tend to hear but we don't). If your dog doesn't currently react to the sound of someone at the door, all you need to do is set up some practice situations with a friend or family member who gives them treats after knocking or ringing the doorbell (as described below), and the dog should learn to show excited behaviour that can be further rewarded. If you want a more specific way of them telling you that someone is at the door, for instance if you are hard of hearing, then backchain a formal alert using the process below. The doorbell is probably the easiest task to train from the dog's point of view, as seeing people is so rewarding (for most dogs).

STAGE 1

Have a helper stand just back from the open front door (being careful that the dog cannot run out into the road) and show your dog that they have some tasty treats. Hold your dog gently by the collar away from the door (but in sight), ask the person to knock or ring the doorbell if you have one, and let go of the collar and let your dog run to the reward. Then practise this from all over the house.

STAGE 2

Progress to having the door closed. The dog now has to wait for you to follow them to the door and open it before they get their reward.

STAGE 3

Once your dog is super confident at running to the door excitedly when cued, don't follow them. Sit reasonably close to the door, and let the dog run to the door and look at it, then prompt them using your alert cue to come tell you about it first. Once they have alerted, you can ask, "What is it?" Some dogs with good memories may run back to the door. Most dogs will then look at you and convey, "I've forgotten!" This is when your observant helper needs to quickly ring the doorbell or knock to remind the dog where they are going, and fling the door open and encourage the dog to them for a reward while you follow.

STAGE 4

From now on, your dog must alert you before you go to the door, and as you practise will need less reminders from your helper at the door, or from you with the alert cue – slowly phase these out. Gradually progress to keeping the door closed until you get there, and alerting you at further distance away in all rooms of the house.

STAGE 5

Once your dog is confident enough to perform the behaviour reliably, you must also switch to the reward coming from you instead when they get there (keep a pot of treats near the door). For real-life doorbell situations then remember to quickly put the dog somewhere safe (such as behind another door near the front door, or in a crate) whilst you open the door to the visitor/delivery.

MEDICATION REMINDER

You may need to take medications at particular times each day, and a persistent dog can help you do that. The same training stages as used for the doorbell can also be used to teach your dog to alert you when a timer or alarm goes off. The only real difference is that your helper switches a timer on (which could be a mobile phone, if needed) instead of ringing the doorbell, and which can be placed anywhere in the house, rather than the specific location that is the front door. Once trained, your dog can take you to the source of the beeping, and if you only ever give them their food reward (or toy) after you have taken your medication, they can be quite persistent in making sure you don't forget them. However, if you want to get really fancy, you could even train your dog to bring you a medication bag or pouch when they hear the alarm.

ALERT FOR DROPPED ITEMS

In the previous chapter, we taught the dog to pick up items we had dropped. We can also add in to alert us and let us know if we drop something. To do this, backchain the alert on to the original task.

STAGE 1

Ask someone to gently hold your dog by the collar for you and ask them to let your dog go when you drop the object very obviously. Your dog will run towards the object. Use your alert cue to encourage your dog to come and alert you instead of going towards the object. Reward your dog for alerting you and then ask, "What is it?" and turn your body towards the dropped object. As soon as your dog looks at the object, reward them. Then move on to encouraging them to pick up the object on your command only (so that you have checked that it is a safe item to pick up). Practise this until your dog runs to alert you before going to the object.

STAGE 2

Begin to practise this with different distances between the object and you, and with you in a variety of positions, until you can walk with your dog next to you, and your dog will stop and alert you and look at what you have "accidentally" dropped.

CLOSING AND OPENING DOORS

Jasmyn had a habit of walking into rooms and leaving the door open, so I taught her to shut it behind her. Similar to how we teach a "touch" to the palm, you can teach a dog to push a door with their nose. Instead of targeting their nose to touch your palm, you teach them to touch a sticky note (at first, place it on the floor with a treat on it). Once they can run and touch the note vigorously, you stick it on a door at their head height, and later add a cue such as "shut the door". Eventually remove the sticky note completely once they are confident in shutting the door (if needed, you can gradually cut it smaller, to fade out the visual clue). Small dogs may need to use their paws rather than their nose, but I recommend they touch with their nose if possible – from experience, I regretted teaching Jas to enjoy launching her front paws at doors in a rented house.

To teach a dog to open the door, it may be as simple as pushing it from the other side. Alternatively, you may need to ask them to tug on a rope which is tied round the handle, in order to pull the door towards them. By now, you should have a good idea of how you might train this – first teach them to tug the rope in your hand, before tying it to the door and rewarding them for pulling it there, and then add a command such as "open". There are also toys that have a ball on one end and a suction cup on the other, and these are handy for sticking on to doors (or even sliding freezer doors in supermarkets) for dogs to open by tugging.

OPENING DOORS BY PULLING A ROPE ON THE DOOR HANDLE

MEDICAL ALERTS

In the two case studies I described in Chapter 1, pet dogs had closely observed their owners and started to help with their owner's health needs. They would approach, comfort and lick them when the owners were anxious, or look at them worriedly and press their nose into their owner's body when they thought their owner was about to experience a medical episode. The dogs had received no specific training for this "alert", and it is common for dogs to start naturally responding to the way their owners behave, in particular because their reaction is often rewarded with attention. However, these particular dogs were perhaps reacting in a worried or anxious manner to their owner's behaviours. Although this may still be helpful for the handler, it would be better for the dogs' welfare if their reactions were more confident and relaxed.

It is crucial that we notice if the dog's reactions are signs of anxiety – pacing, whining, nudging or even barking ("Oh no, she's doing that weird thing again, I don't like it, please stop, Mum!"). Far better is if the cue to the alert behaviour is paired with something positive that the dog enjoys, and so the dog is less likely to be anxious when it happens. ("Yay, that thing is happening again – this means I get my ball!"). Ideally, further training could solve this, and also help the dogs to be more clear and specific in their helpful behaviours.

Dr Claire Guest is the CEO and founder of the charity Medical Detection Dogs (MDD), and tells me:

> "Medical alert assistance dogs are actually trained to recognize the odour change normally caused by metabolic and hormonal processes in their owner

that precede an oncoming medical emergency. Dogs, having highly sensitive and sophisticated olfactory apparatus (noses), can easily identify these odour changes and our companions are highly tuned to understanding that these changes signal a threat. For many acute medical episodes, although a behavioural change in the owner may occur, this is often too late – for example, rapid drops in hypoglycaemic episodes in diabetics, Addison's crisis or postural tachycardia syndrome (abnormal increase in heart rate). Therefore, the charity MDD trains dogs to identify odour signals to ensure an early alert, which enables the client to take medication and make themselves safe."

Training an alert for signalling medical episodes through detection of changes in odour is complex, so I advise you to seek specialist support.

IDEAS FOR MENTAL HEALTH TASKS

Unlike "emotional support dogs" that supposedly make people feel better through their mere presence, dogs can also be trained to do physical tasks that help us greatly with our mental health. Tasks we have already covered are the alarm clock, touching or cuddling, and medication reminders. Other ideas are to "block" people from standing close to you, either by standing or sitting in front of you when people approach, or standing or sitting behind you. You could simply lure your dog into position at first using your hand and some treats, then add a command such as "in front" or "go behind", and practise until your dog can follow the command without the treats in your hand luring them there. Alternatively, you could use a helper approaching you from the front or from behind as the cue for the behaviour, rather than you needing to tell the dog what to do. This means they can learn to do it when you don't even realize a person is approaching.

Another useful task is to interrupt you when you are doing a behaviour you wish to stop. This may be things like spending too long at the computer arguing with people on internet forums, staring at your phone, or an extreme repetitive behaviour such as washing your hands, scratching, or chewing your nails. At first, practise calling your dog over to you while you are acting out that behaviour, and praise or fuss, play games with their favourite toy, or give treats. Once they have practised this a few times, try acting out the behaviour without calling them and see if they approach (if not, give a little extra encouragement). You could use your formal alert in this situation (or not, if you need that for another specific thing only – in this case, just teach them that any general pestering works). Be careful to remember to actually respond to them in real scenarios too, otherwise they may learn to give up quickly as they are being ignored. Lots of practice using "fake" acting of these behaviours can help prevent this.

CAN I TAKE MY DOG INTO PUBLIC PLACES?

I've given you examples of many things that your dog can now do to help your health and well-being. Does this mean that you can now take it with you into shops and restaurants? The short answer is no – unless you have a recognized disability, for which the law then typically allows reasonable adjustments, including the presence of your trained assistance dog. However, even if this was the case, I would still suggest you don't. Performing physical tasks is only a small part of the vast training and preparation that an assistance dog receives for working life. Most of my time as an instructor was spent training the dog to cope with – and behave impeccably (without distraction) in – shops, malls, restaurants, elevators, and public transport, including passing a Public Access Test. Assistance dogs also go through months of intensive socialization programmes, and come from specialized breeding programmes, all to prepare them specifically for these types of situations.

If you feel you desperately need this level of support from your dog, there are some organizations that can support you to train your own dog, and hopefully certify your dog as meeting the high standards required for public access. I cannot stress enough that training your own dog without some form of external support is going to be extremely hard work and likely not in the best interests of your needs or that of your dog. An alternative is you apply for the waiting list for an organization-trained dog. Either way, please research your potential organization thoroughly.[7] If it seems easy to get certified, or they ask you for money, they may not have your welfare or that of your dog at the centre of their work.

TEACHING A SETTLE

Thankfully cafés, pubs and even some shops are now becoming increasingly "dog friendly" for even pet dogs. Therefore it is a good idea to teach your dog to be well behaved in these environments, so that business owners don't change their minds due to unruly behaviour, and even more are encouraged to allow us to bring our pets. Rather than just expecting your dog to know what to do if you take it to the pub, you need to teach it first what it means to "settle".

> **Note**
>
> If you have more than one dog, this exercise should be done either with both dogs at the same time, but each with a different person, or with the other dog shut out of the room.

STAGE 1

Start at home at a time when your dog would normally be quite settled. Ensure that [s]he has recently been outside to empty herself. Put your dog on her normal lead. Tie the lead to the chair where you are sitting, or if that is not possible, put your foot on the lead, nearer to your dog's collar. The lead should be fairly short, so that your dog cannot tie herself in knots and so that it is comfortable for herself to lie down. You are going to stay with her; you must not move away. Lure your dog into a down position. If your dog does not like lying down on hard floors, you may want to lay down a small piece of blanket for her first. With your dog lying down, tell her to "settle" (or whatever command you choose).

STAGE 2

Ignore your dog for a few minutes. Most dogs settle quickly, but be patient. If your dog struggles, paws at you, whines, etc., simply ignore her. She must not learn that he gets more attention for misbehaving. If your dog absolutely cannot cope with being ignored and is freaking out, rethink the exercise and give her something safe and tasty to chew on while she is lying there instead.

STAGE 3

When your dog is lying nicely and quietly, and only at this time, you can quietly tell her she is being good and give her a food reward. It is best to drop the treat on the floor rather than give it from your hand, so that she is encouraged to stay lying down and not get up. If your dog prefers to quietly sit, that is OK, but lying down is preferable and she is likely to be more genuinely relaxed that way.

STAGE 4

Increase the length of time for which your dog is being asked to settle. If you need to get up and go out of the room, you must release your dog first. You must not release your dog unless she is settled. When the time comes to release her, she must know that the exercise is finished – so use your normal release command that she already knows. She needs to understand when you want her to settle and when she can move, so that he doesn't keep trying to see if she can wander around during the exercise.

STAGE 5

Progress to doing this in other rooms in the house and when there are distractions (you are eating a meal, friends are visiting, or you are in the pub), or anywhere you want your dog to settle down quietly.

CHAPTER 13

HELP, I HAVE A PROBLEM

DEALING WITH BEHAVIOURAL ISSUES

Pets aren't perfectly behaved all the time – nor, I suspect, would we want them to be. Erica Peachey has found a humorous solution to the frustrations our pets can sometimes produce. She assigns a "Troublesome Pet of the Day Award" to individuals in her plethora of animals, when needed – at least they are being "good" at something that day! However, sometimes our pets' behaviour is more than an inconvenience and funny story to tell, and becomes a serious and stressful problem – to us, at least (it may not be to the animal).

Prevention is easier than cure. By following the guidance in this book, hopefully problematic situations will not present themselves. But life happens, and even a previously confident dog may develop a sudden fear of something or be traumatized by an event, such as being scared by a boisterous dog. Serious behaviour problems such as aggression will damage the dog–owner bond and may make you less likely to take care of your dog's needs, such as exercise. Aggression to strangers and to other dogs in particular will make taking your dog for a walk far more difficult for you, and thus reduce the motivation to do it. Even relatively minor unwanted behaviours may make you like your dog less. For these reasons, it was important that I dedicate a chapter to this topic, but one book chapter cannot solve what are often complex underlying causes and varied behavioural symptoms. I do aim to give you an idea of the kinds of approaches to take when addressing unwanted behaviours, and how to get support.

MISBEHAVIOUR

When dogs misbehave, the temptation is to administer some kind of punishment. If the misbehaviour is repeated, or we are in a bad mood for some other reason, the temptation becomes stronger. However, if you have the correct relationship with your dog, you can assume that he does not know *how* to do the right thing, or he would be doing it. Always remember, if you tell your dog off more than three times for the same thing, it possibly isn't working (see overleaf). Look at the situation from a different angle, work out why he is misbehaving and approach it in a new way.

As with children, there tend to be one or more stages where the dog seems to have forgotten everything he has ever learned, he does not do as he is told and thinks of all kinds of "naughty" things to do. The most usual signs are that the dog doesn't come as quickly when he is called, and when he does come, it may be difficult to catch him. He appears to have forgotten everything you have taught him and sometimes doesn't respond to his name. He may be slightly pushier around the house, as though he never knew the rules you taught him a couple of months ago.

Congratulations, you have produced a normal, happy, well-adjusted and probably adolescent dog.[1] He is now simply testing to see what the words mean and what the rules are. He needs to know: does "sit" really mean "sit now", or does it mean "it doesn't matter if I don't bother to sit". Be patient, this phase will pass. You must now be even more consistent,

---

IS THE PUNISHMENT WORKING?

If you do ever punish your dog, you must be certain that:

1. **He understands what it is for.** When you return home, there is no point in telling him off for something he chewed an hour ago, or for not coming when you called him when he does eventually return to you. He can only connect the last thing he did with any punishment or reward.

2. **It will reduce the likelihood of him doing it again.** Telling a dog off for being nervous will not make him any less nervous next time. Telling a dog off for jumping up can make him more frantic to apologize to you. If his way of apologizing is to jump up to lick, you could be making things worse.

3. **You are sure that it is doing some good** and not damaging your relationship with him.

4. **You are not simply telling him off** because you think it will make you feel better or because you are losing your temper or annoyed about something completely different.

and ensure that your dog is learning what you want him to. Rewards for the right things are even more important. Thinking that you need to resort to punishment during this period is a bad idea, as dogs going through adolescence are likely to be particularly sensitive at this time.

Any time that your dog is "being difficult", ensure that you and your family are still being consistent with him. Have you lapsed with the rules you taught him at first? Are you still spending time teaching him things? Do you still put aside some time for playing games with him? Are you still noticing what he is learning and rewarding the good things? Or have you begun to take his good behaviour for granted and feel that he no longer needs rewards, as he should know by now? Unfortunately, if this is happening, a dog will become confused and unwanted behaviour is likely to develop.

Before assuming that your dog is misbehaving, check that you are still behaving in the right way towards him.

WHEN TO SEEK PROFESSIONAL ADVICE AND FROM WHOM

If your dog develops a significant behaviour problem, there is a primary question to be asked – is the cause medical? Highly qualified dog behaviourists and trainers only work on veterinary referral for this reason. If your dog's behaviour changes suddenly, it may be caused by an underlying medical condition that is causing her pain or other problems, such as digestive issues. Therefore, seek advice and a full check-up with your vet first. If that doesn't find a solution, you likely need to seek advice from a dog behaviour expert. Professor Daniel Mills is a veterinary behaviourist whose clients are often referred to him after having a veterinary check, and he thinks that even then, the role of pain often goes unrecognized.[2] He tells me:

> "Indeed, at our referral clinic, in excess of 80 per cent of problem behaviour cases seen by us have some form of association with a medical and often painful condition. There are many ways in which pain can affect dog behaviour and I hope to increase awareness of the problem."

According to Daniel, the association between pain and behaviour can be a direct manifestation – often the dog has a generally poor temperament, is reluctant to move, is aggressive in short bursts towards a number of people, and directs that aggression at people's hands when they try to touch the dog. But often the association can be less clear, with pain exacerbating an established behaviour problem, or causing a specific symptom of it:

> "One example of this is a case I treated where the dog had a persistent digging behaviour when left alone, likely caused by pain in its leg, which it associated with lying down when left alone. It was only when the cause of pain was treated that the dog's reaction to being left alone substantially improved, alongside the training plan of course. In cases such as these, where the training appears to not be working as expected, perhaps a course of pain relief for 4–6 weeks may elucidate whether pain is involved."

Another situation for which I would recommend that you seek professional advice is if your dog won't take any food treats that you try to use in training. There is a likely cause for being unmotivated by food, and it may be that your dog is still feeling too stressed out in that situation to think about eating. They may also have associated food rewards with bad things happening. An experienced trainer or behaviourist can help you teach your dog to enjoy eating and use it as a reward in training. Without overcoming this issue first, you are unlikely to make much progress.

HOW TO FIND A BEHAVIOURIST

Do you need a dog trainer or behaviourist? Some professionals do both roles, but in general a dog behaviour specialist will deal with more serious issues (phobias or aggression) than what could be considered training matters (pulling on the lead or not sitting when asked). (I add a caveat here – even these behaviours might actually have more serious underlying motivations, such as fear or pain.) Training is more about teaching the dog to do something, whereas changing behaviour is about looking at how to stop the dog doing something, and often involves changing their emotions. In Chapter 7, we discussed how to choose a dog trainer, and the same issues apply here when looking for a behaviourist. Note that you should avoid anyone using punishment- or dominance-based methods, as these will likely make the problem worse, even if it seems like a quick fix at first. Again, thoroughly research the calibre of the person's qualifications to discern whether they are modern, scientific and formally assessed. Membership of a reputable body means that they are answerable to someone, and should indicate a higher quality of work.[3]

Whether working on the issue by yourself or with the guidance of a behaviourist (recommended), the fundamental approach taken should be to gradually get the dog used to whatever the problem situation may be. This includes changing the dog's emotion in that situation (usually from worried to relaxed and happy), and finally we teach the dog a new and more acceptable behaviour to perform in that situation instead. Many people ask me "Why does my dog do X?" My response is typically "Why not?"

PRINCIPLES OF TREATING SEPARATION PROBLEMS

Soon after I rescued Jasmyn, the two of us moved hundreds of miles away from home in order for me to start a new job training as an assistance-dog instructor. At first, we lived in a studio apartment, with a shared bathroom down the corridor. Unfortunately, Jasmyn was showing quite extreme separation behaviour after coming out of the shelter kennels, and I could not even leave her in my room while I went to the toilet without a barking frenzy, let alone while I went out to try to make new friends. For Jas, the key to treating her separation behaviour was to make her feel safe and secure. To do this, I used a crate, which I gradually let her get used to being shut inside for short periods, with a safe chew toy to occupy her. At the beginning of this training, she was so stressed that she even ignored the food (despite being highly food motivated), which is not unusual. Imagine you are about to do your school exams; are you hungry? It required a lot of incredibly gradual work, first getting her used to being in the crate and eating with me there, and then only briefly going out of her sight for mere seconds, before she was comfortable enough to let me go to the bathroom without her. For Jas, covering the crate so that she couldn't see out and anticipate for my return seemed to help her settle. Over the years, she progressed to being left in just one room instead of a crate. Eventually she was comfortable being left with the run of the house, as long as she couldn't "watch" for me through the front windows, which always seemed to unsettle her, so I kept her to the back of the house for that reason.

Sidney was one of the first separation-related cases that I treated as a behaviourist, and after all my work with Jas, I was confident I could help his owners. However, nothing from the textbooks or my experience with Jas appeared to work. One day, in desperation, the owner risked leaving him with the run of the house instead of a crate in the kitchen, and videoed his response – he curled up on the owner's bed and went to sleep. I learned a key lesson from that case: every dog is different. This is one reason why it is important to work with a qualified behaviourist, who not only has their own knowledge, but the combined expertise of an entire membership organization of behaviourists to draw upon when needed.

PRINCIPLES OF TREATING FEARS AND NERVOUSNESS

Brie was quite frankly a nervous wreck when she first arrived with us, as we were told she lived her first few years of life in a puppy farm. For days, she hid under a bush in the garden, and I had to keep a trailing lead on her so that I could gently retrieve her into the house when needed without touching her. Yet now when people meet her, and she inspects them for treats and strokes, most would never guess what she used to be like. It took many years of gentle exposure and socialization through walks, household visitors and going to training classes, through which she learned that people were not a threat (and may also have something tasty for her).

All dogs become frightened by certain things. You should have a good idea of what may cause your dog to be anxious. Your aim should be to teach her gradually not to be frightened. This takes time and patience, and its success relies on you knowing your dog and being able to read the signs that tell you when she is confident and when she is apprehensive, and changing your behaviour accordingly.

For example, if she is frightened of the vacuum cleaner, play with her and feed her when it is near, and this will help to teach her not to be worried. If necessary, progress extremely gradually. Teach her to be happy in the same room when it is not switched on, then teach her to get used to the sound when it is in another room, and then you can gradually move her nearer. Do not go too quickly, which could frighten her. The aim is to increase her confidence and ensure that nothing frightens her and sets back her progress. Also be careful not to make the food reward you are giving her "predict" a scary thing, or you may accidentally teach your dog to be suspicious of food. The food reward should come after they have seen, or made some attempt to investigate, the scary object or person.

If something does happen that is out of your control (like a large lorry speeding past when you are out, which frightens your dog), do not throw your arms around her or make a big fuss of her. Although this reassurance is natural for humans, this is confusing for dogs and may make her more nervous in the future. Instead, act confident and distract her, perhaps telling her to sit and rewarding the sit, playing with her as a reward for walking away, or anything else which takes her away from the worry – and then you can reward her confidence.

If you know that your dog has a difficulty that results in fear, or is going to have a stressful time, such as during fireworks, consider seeking professional help.

PRINCIPLES OF TREATING AGGRESSION TOWARDS OTHER DOGS

Aggression towards other dogs (outside the home) is a common problem. Some dogs are reactive to all dogs they see, but others only react to dogs of a certain breed or colour. This may be due to generalizing, for example, all black dogs after a bad encounter with one. It may also be because the way we have bred some dogs to look (or modified their ears or tails through surgery) can make it difficult for other dogs to read their body language and know their intentions as friend or foe. Aggression can be caused by fear, but also by frustration at not being able to interact with a dog they see (especially if on a lead). Professional advice regarding the correct motivation for your dog is important.

Jasmyn had previously been fine with other dogs. She was never the most social animal, but she was usually happy to sniff and have a little play with most dogs we met. However, she gradually started to freeze and then snap at dogs after these brief interactions, usually at the point where the dogs would otherwise have gone their separate ways. This was obviously not acceptable. So every time Jasmyn started to sniff and greet a dog, I called her away after a few seconds, before the snapping could occur. When she came to me, I rewarded her with a treat. Over time, she began coming away from dogs of her own accord and looking at me expectantly for her treat. She even reached the point of deliberately approaching dogs for a sniff, looking at me to check that I was watching and then coming to me and asking for her treat.

Ben the collie later developed a similar problem. He used to get picked on by male dogs who, for some reason, liked to sniff his bottom a lot and got sexually excited by him. Eventually, even timid Ben decided to defend himself and started to approach dogs first, circling them, and then giving a little air snap in their direction. Ben was not motivated by food, but he loved a ball. He quickly learned that whenever he saw a dog, I would produce the tennis ball from my pocket and throw it for him. This meant he could go back to ignoring dogs and focusing on me instead. His emotion around strange dogs also changed, as he naturally came to me and sat to wait for his ball – and a bottom on the ground is far less tempting for another dog to investigate!

Essentially the same problem in these two dogs – getting snarky with other dogs – was solved by doing the same thing but with a different motivator, either food or a ball. It was also successful because I intervened early, before either dog became very fearful of other dogs, which allowed me to fairly easily change their emotional reaction and behaviour upon seeing another dog.

PRINCIPLES OF TREATING AGGRESSION TOWARDS PEOPLE

Aggression to people is a serious behaviour problem that certainly requires specialist help. The underlying causes may not be apparent from the behavioural symptoms, and a full behavioural history will be taken. Whatever the cause, the general approach will be the same as above, but nuanced for that particular dog in that particular context. Typically, the dog is fearful or nervous, often due to a past experience, and the more subtle signs of their emotional state have been ignored until they have learned to respond aggressively. A careful and controlled training programme is required in order for the dog to learn new behaviours in this context, and for them to feel different emotions too.

If your dog is showing aggression towards people, it is highly recommended that the dog learns to wear a muzzle, at least in the short term. Actually, it is helpful for any dog to be trained to be happy to wear a muzzle, in case of a future situation in which they may need to wear one, such as if they are in extreme pain at the vet.

MUZZLE TRAINING

Muzzles can help when overcoming a whole range of behaviour problems. Although some people have negative views towards muzzles, or rather towards dogs wearing them, they do have their uses. They will not change the way the dog wants to behave and therefore they should not simply be used to stop unwanted behaviour. They can give great peace of mind, as less damage can be inflicted. However, it is unfair to simply put the muzzle on the dog and hope it will be all right, or worse still, to use the muzzle as punishment for the dog. Time and care must be taken to ensure the dog likes wearing the muzzle (overleaf).

There are two kinds of muzzles. Firstly, the soft fabric muzzle, which fits like an open-ended sleeve over the dog's

STEP-BY-STEP MUZZLE TRAINING

RULE: Whenever the muzzle is near your dog, tell him how wonderful he is. When you take the muzzle away from him, switch off, put the treats away and ignore him for a little while.

Repeat each step many times until you know that your dog is completely happy, before progressing to the next step:

1. First of all, let the dog see, sniff and take tiny pieces of treats that you have placed in the entrance to the muzzle.

2. Put really tasty treats in the end of the muzzle and let the dog put his nose right in and get them out.

3. When he is happy putting his nose in, gently slide the straps around his ears while he is eating the treats, but do not fasten the muzzle. (If you are using a basket muzzle, you can put something sticky, such as cream cheese or peanut butter, at the end.) If your dog is backing away, do not try to do this – keep working on Step 2 until your dog does not back away from the muzzle.

4. Ensure you have some more treats that he can eat with the muzzle on. Put the muzzle on as above, fasten the clip and feed him some extra treats before taking it off and then ignoring him.

5. Repeat this for as many short periods of time as possible, gradually reducing the number of treats he eats and increasing the time the muzzle is on.

6. When your dog is happy to wear the muzzle because he knows it gives him your attention and treats, you can extend the time even further and try encouraging him to walk around with it on.

7. If at any time the dog tries to scratch at the muzzle, try to distract him and reward him for stopping. This means that you are trying to progress too quickly.

8. Now that your dog is happy wearing the muzzle, it means you can put the muzzle away and use it at times when it will give you greater peace of mind. Any time you are a little unsure, you can put the muzzle on your dog. Because you are now less worried, and the dog likes wearing the muzzle, the dog will find it easier to relax.

It is normal to take at least two or three weeks to progress through this plan. In order to avoid your dog associating the muzzle with stressful situations, sometimes put it on when nothing is happening, so that your dog accepts this as normal.

nose. This has the advantage of being less obtrusive, and the dogs often learn to accept it more quickly. However, the dog cannot pant with this on and so it is unsuitable for dogs that need to pant, for use over long periods or when exercising the dog. It is also possible for dogs to still nip through the gap at the end. Generally, these muzzles are of little practical use to owners. The other kind is the basket muzzle, which is a lightweight, plastic, box-type muzzle. This fits over the dog's nose and allows the dog to open his mouth, pant and drink while wearing the muzzle. Because these are rigid, some dogs are slower to accept them, but they are much safer for people and promote a higher level of welfare for the dog.

CHAPTER 14

PETS AND PANDEMICS

(AND OTHER DISASTERS)

I began working from home full-time a few weeks before the first UK national lockdown of 2020; working in a department that studies infectious diseases, including coronaviruses, meant that we knew the seriousness of what was coming and transferred to homeworking and remote teaching fast. It wasn't long before my husband did the same, and then schools closed, too. Suddenly, we were with the dogs all day, every day. After a few weeks, Roxie started barking and yipping in the middle of the night. I came down and let her out, in case she needed to toilet, but she just stared at me. This happened a few times. Eventually, I came to the conclusion that she was just trying to get our attention, in particular that of my husband, whom she had been spending a lot of time sitting with, not only in the evenings as usual but now all day when he was working. (For some reason, the dogs rarely come to see me when I am working – I would be offended, but perhaps it's a sign of a healthier relationship.) Over the next few weeks, we actively worked on the problem, making sure that he didn't let Roxie stay with him all the time, shutting the door to where he was working; the barking stopped.

As well as turning life as we know it upside down, the 2020–2021 COVID-19 pandemic (during which I am writing this book) has also put a fascinating spotlight on our relationships with dogs. So many people got new puppies! I love the fact that when re-evaluating our lives during this time, many decided that going forward, it had to involve a canine companion. But how we interact with the dogs we already have has drastically changed too. I'm pretty sure Roxie and Brie are loving the "new normal" and never want to go back to their old lives. Stay-at-home orders, social distancing measures coupled with working from home or being furloughed, and limits put on outdoor exercise (such as UK stipulations of once a day) truly changed our routines with our dogs. Research confirms that the pandemic led us to spend much more time with them: an online survey by the charity Dogs Trust showed that just over half of dogs were not left alone for more than five minutes a day (compared to one in seven before the lockdown).[1] Seventy per cent of dogs were spending more time with adults than they were before the pandemic, and even more were spending more time with children than they were used to.

IS ALL THIS EXTRA ATTENTION GOOD FOR DOGS?

A handful of owners in the Dogs Trust survey reported dogs being overwhelmed with the increased attention and said that their dogs reacted aggressively towards them or their children. Likewise, colleagues and I are currently analyzing data from our local children's hospital, which shows that Accident and Emergency attendances for dog bites soared when children were off school, even when numbers of children visiting A&E for other injuries or illnesses were incredibly low compared to normal. This is worrying, and something to bear in mind during inevitable future lockdown situations.

A more common reaction was dogs becoming more clingy and attention-seeking at home (nearly half), and some started showing separation-related behaviours like Roxie did. Being able to spend some time alone now would help dogs to cope with being alone when owners return to work. The research by Dogs Trust found that although most dog owners noticed separation-related behaviours in their dogs and worried how their dogs were going to adapt to being alone again, unfortunately, owners rarely implemented any measures to prevent future behaviour issues, such as I have been doing with my dogs. The study suggested that owners find it hard to leave a dog in a separate room when the whole family is working or studying at home. However, it is better for the dog's welfare to address these issues proactively. Our dogs now seem to be coping much better when we are in the house, happily taking themselves off to other rooms to snooze. Still, I have noticed that when one person goes out, the dogs are more watchful than before, waiting for their return and becoming overexcited when it happens. For this reason, we are making sure that as much as we can, people come and go, including all of us leaving the dogs alone in the house sometimes for a short while, just so that they don't get so used to someone always being here. The problem for most owners (including us) is that having a dog nearby may be just too wonderful a thing to want to change it.

DOGS HELPING US COPE WITH
"UNPRECEDENTED TIMES"

Unsurprisingly, our mental well-being deteriorated during this period, with 38 per cent of UK adults feeling significant anxiety and/or depression in April 2020 (up by 13.5 per cent compared to previous years).[2] However, a UK study found that pet owners perceived their pets as a source of support, and had a smaller decrease in mental health during the first lockdown phase compared to those who didn't have pets, regardless of animal species.[3] Research in Australia found that having a dog in particular protected against experiencing loneliness during the pandemic.[4]

During the early months of the pandemic, myself and colleagues conducted an online survey of over 500 UK dog owners and found that dogs offered companionship, routine and structure to the day that provided a semblance of normality in a drastically non-normal world (currently unpublished.) Their responses show just how integral our dogs are to our mental and physical health:

"[My dog] is the sole reason I get out of bed in the morning."

"Everything else in our lives has changed; however, the need to care for our dog remains constant."

Walking a dog was often seen as the highlight of the day. It provided an opportunity to get out of the house, and to be present in the moment rather than worrying:

"I probably wouldn't have gone out at all, other than to the shops, if it weren't for the dog."

"Allowed me to get out of the house and escape the family."

"Dog walking is the best thing in the world to unwind and be in the moment."

Being able to pet and touch dogs when physical contact with people was not allowed was a lifesaver for many owners. Dog walking also allowed us to socialize with people (albeit at a distance):

"Can't hug or kiss or touch other people outside the household so dog is an extra cuddle."

"Felt less isolated as saw other dog walkers I knew when out and about."

We also found that dogs provided daily amusement and laughter, a welcome break from nonstop reports of rising cases, which owners welcomed:

"I have somewhere to focus my energy that's not on news and bickering and the terrifying death toll."

"Brought fun and laughter to the household [during] otherwise depressing time."

"A nice topic to talk to family and friends about."

Although caring for a dog during the pandemic wasn't stress-free – with many of us worrying about being able to meet dogs' needs, access pet supplies and receive veterinary care (which was often "emergency only")[5] – research has demonstrated that the value of their companionship largely trumped these worries.

CHANGES TO DOG-WALKING ROUTINES

The COVID-19-related restrictions also impacted how we walked dogs. When asked what their opinions were regarding the change in frequency of dog walks (total number of walks during the week) during the first UK lockdown, half of our survey respondents (52.8 per cent) reported no difference; just over a quarter reported a decrease; and even less reported an increase. Overall, we found that during lockdown, dogs were on average walked less often but for a similar total duration per week, meaning that (at least for most dogs) the walks were made longer to compensate. However, the research data collected and analyzed by Dogs Trust disagreed with our findings and suggested that dogs were walked less often and for less time daily (although the decrease for the latter was less drastic). In Spain, where lockdown enforcement was strict and well defined, only one person at a time was allowed to walk the dog, who was to be kept on-lead at all times, potentially further restricting the dog's ability to exercise and explore the environment; over half of owners felt that their dog's quality of life had got worse.[6]

We found that in the UK, people whose dogs usually received multiple walks per day coped by having different household members take them out each time, but single-person households with high-energy dogs suffered. Owners who considered themselves vulnerable and tried to isolate at home also reduced the duration of their dog walking. The Dogs Trust research also found that the reduction in daily frequency of walking was associated with respondents living in villages or cities (compared to rural/remote locations), being unable to walk their dog off-lead near home, and having a young dog (under two years of age). Whereas previously, some people relied on the help of dog walkers, we found that duty fell on the owner and household members instead, especially if they were now working from home or furloughed (which I can't help but feel is a positive outcome to emerge from it all).

For some, walking under the lockdown restrictions was also somewhat stressful and less enjoyable than before. Both surveys agreed that dogs were more often kept on leads and allowed to socialize with other people and dogs less often than before. We found that owners often changed where they walked their dog to be able to stay socially distanced and to avoid travelling for walks. Instead, dogs were walked near where they lived, on streets, and in contrast, travelling to walk dogs in places like beaches, woodlands or country parks decreased. Compared to my previous research, these changes may mean that walks became more "functional", and less "recreational" in nature. But given how important our daily outdoor exercise has become to us when we are stuck at home, it could be argued that even these more "boring" walks had great value to the owner and could still be considered recreational.

The COVID-19 pandemic was also a time of increased conflict between users of public spaces, both between

dog owners and those without a dog, and between dog
walkers themselves:

> "The most stressful part of dog walks are all the extra
> families with children who have no dog/outdoor
> sense, and a general lack of awareness and don't/won't
> follow social distance rules. They seem to think they
> have priority over everyone else just because they have
> a child, even though they don't normally use any of
> the outdoor spaces like the rest of us (dog walkers/
> riders/runners/cyclists) do."

> "Far more people out walking their dogs, who frankly
> don't have a clue about walking their dog. Dogs
> off-lead, no recall. Walking with headphones on or on
> their mobile phones with no awareness of their dog."

Given what a stressful time in general this was (and still is),
perhaps it's not surprising that frustrations were directed at
each other. Although many owners found their usual walking
places more crowded and full of annoying "newbies", leading to speculation that people were suddenly walking dogs
that didn't usually get walked, our other findings suggest
it is actually just that they were previously walked in other
places. Some owners struggled to socially distance in these
crowded spaces, and to stop people touching their dogs, leading to fears of their dog catching the virus on their fur. This
has been a particular issue with Roxie, who usually attracts
a lot of petting. She can't understand why some people now
don't want to say hello, and is confused as to why we keep
calling her away from other people who are trying to get her

to come over for a stroke. For some owners of anxious or dog-reactive dogs, walking in these crowded spaces became a nightmare. However, other owners of reactive dogs reported social distancing being helpful as people and dogs now avoided each other more, and some places became quieter. Owners reported that some dogs have changed in their behaviour, possibly due to reduced socialization with other dogs and people; they became more reactive, nervous or exuberant when meeting people or dogs.

Thus, it certainly wasn't "business as usual" for dog walking during the pandemic. But what if the owner actually caught the virus? Of our 1 in 10 survey respondents who thought they had had COVID-19, or even more who thought a household member had it, over a third continued to walk while they were supposed to be isolating at home. Reasons for continuing to walk included: not having anyone else to help; not trusting anyone else to help because dog is reactive; living in rural areas or having access to own private land where they never see other people; presented symptoms early on in February, when little was known about the virus; and feeling it was important for their mental health, so long as extra precautions were taken (e.g. walking early in the morning or late at night to avoid others).

PANDEMIC PUPPIES

I know a lot of people who either got a new dog in 2020 or are looking for one. With so many people working from home, lockdown seems the perfect time to spend time raising a new family member. Studies have shown that both rescue-dog adoptions and online adverts for new puppies have increased dramatically.[7] Not surprisingly, the demand for puppies and rescue dogs has been reflected in increasing prices for those wanting to get a dog.[8] For example, in March before the lockdown, an average price of a dachshund was £973, compared to £1,883 by June 2020.[9] In late 2020, my friend Debra was looking for her first dog and being asked for around £3,000 even for a cross-breed pup. The demand has been particularly high for breeds most commonly smuggled to the UK after being bred in puppy mills abroad (dachshunds, English bulldogs, French bulldogs, pugs and chow chows). We have already discussed that puppies bred in large breeding schemes such as "puppy farms" (whether at home or abroad) are more likely to suffer from behaviour and health issues later in life due to lack of care and poor socialization. Commercial breeding establishments have ramped up their production to meet the demand, and owners of family pets have also decided to cash in and have a litter, with little knowledge of the complexities involved in raising genetically and emotionally healthy pups. Buyers may not have the knowledge to recognize poor practice, or their desperate desire outweighs their concerns. UK Kennel Club research found that a quarter of buyers paid without seeing a pup and nearly half didn't see the puppy's environment; nearly a quarter believed in hindsight that they may have accidentally bought a puppy bred on a puppy farm.[10] Thankfully, Debra has me to help her avoid these traps but found it incredibly difficult to find a good breeder, probably because many responsible ones chose not to breed while socializing the pups effectively is so difficult.

For the lockdown puppies, this difficult start is further compounded by restricted ability to socialize them once they are with the new owners. Closure of puppy training classes, and delays in vaccinations needed to allow a puppy to interact with other dogs (as vets are closed for anything but emergency treatments), means finding opportunities for socialization has been even harder. My close friends Dan and Julie picked up a sweet little Cavachon, Amber, in early March 2020. Between her first and second vaccinations, the pandemic's first full lockdown hit, and her second injection got cancelled. Dan tried all the local veterinary practices, to no avail. It was a few weeks later before he managed to find a practice willing to complete her vaccination protocol. During that critical socialization period when they first got her, poor Amber got a really bad deal. Dan tried taking her for car journeys, carrying her on walks into town near shops, and once fully vaccinated, walking her as much as possible so that she could at least *see* other people and dogs if not actually meet them, but the result has not been great. Further restrictions on movement, social distancing and fear of contracting COVID-19 by touching dogs' fur mean that many people are hesitant about interacting with other people's dogs. Plus, new owners such as Dan and Julie have rarely been able to even have visits from friends and family, or visit them.

Amber is now nearly a year old, and I've managed a few socially distanced walks with her. I have noticed she is quite shy and avoids people and other dogs as much as she can. She is slowly improving, and much better with people that she met when she was very young or got to know in between lockdown periods, and she will now approach me to take a treat I toss for her. However, she is never going to be the super-confident dog one would hope for. We will have to wait for wider research on the effect of the pandemic on dogs and particularly puppies born during this time, but if case studies like Amber are anything to go by, it doesn't look good.

2020 may have been the year of pandemic puppies, but all of these issues with getting a new dog in a time like this are a perfect storm. My behaviourist colleagues have many stories of owners desperately seeking help with nervous and even aggressive young dogs, some of which are already seeking new homes. London behaviourist Penaran Higgs (whom we met in Chapter 5) tells me:

"I am seeing a huge influx of dogs who have a multitude of issues. Overattachment to the primary carer is common – possibly as a result of there being constant companionship. This is coupled with owners who either don't realize that quality online puppy training classes are available during this pandemic, or who choose to 'muddle through' using ad-hoc online resources of varying quality, resulting in a bond between owner and dog that is not quite right. I am also seeing dogs who find it difficult to cope when visitors enter the house, because there have been months where this simply wasn't happening, and because the socialization hasn't been spot on. Some of the dogs have generalized anxiety, either through poor genetics, inadequate socialization, or again, a lack of solid and ongoing input from a properly qualified canine professional, so that the small puppy worries have now become huge dog issues. This manifests either in the dogs becoming agoraphobic (the 'flight' fear response) or displaying aggression (the 'fight' fear response) towards their owners, especially once the puppy stage is over."

Add to this that a fifth of those who bought a puppy during the pandemic admitted being unsure about how they will look after their puppy when they go back to work, according to the UK Kennel Club research. Shelters worry that when owners return to work, many more of these puppies will be relinquished. Unfortunately, we are starting to see this happen, with lots of young dogs appearing on internet sites with their owners claiming that they now don't have time for them and trying to recoup the high prices they paid for their pups a few months ago.[11]

It is great for the future of dog ownership that so many people felt they wanted a dog in their lives during 2020 (and beyond). However, if you do get a new dog during this time, you're going to have to make extra and creative efforts to maximize socialization opportunities within these constraints (for suggestions see the Association of Pet Behaviour Counsellors' resources[12] and Puppy Socialization "Bingo", opposite). Personally, I would be very cautious about getting a dog at this time, for the many reasons outlined in this book. It is incredibly challenging to source, socialize and train a happy dog right now, and we know that an unhappy dog leads to an unhappy owner. If you've already taken the leap and have a new dog, going the extra mile now will save you and your dog a huge amount of heartache, both now and later in their lives.

PUPPY SOCIALIZATION "BINGO"[13]

Use food, fun and calm praise when introducing these new experiences. Play the bingo game at your puppy's pace. Keep any training short – no more than five minutes at a time.

Take your puppy to the park and watch others from afar	Dress up in a costume or wear different items (incl. masks) daily	When your puppy spots a dog, give puppy a treat	Hear/see household items (e.g. vacuum, broom)	Gently introduce ear/ teeth checks	Arrange for neighbours to say "hi" from a distance	Create a puppy safe sniffing/ digging spot in the garden	Play hide and seek
Teach puppy to settle on their bed	Listen to recordings of people chatting	See at least 5 different animals from afar	Scatter puppy's food about the garden	Encourage family members to gently handle puppy	Relax with your puppy	See/hear dogs on the TV	Hold a treat high above puppy's head. Wait for 4 paws to be on the floor, then praise and rest
Listen to the washing machine/ dishwasher	Ride in the car at least 5 times	Spot 5 different aged people (e.g. baby, toddler, child, teen, adult, pensioner)	Disappear for 3 seconds as puppy eats. Over time, slowly increase duration	Listen to children laughing/ screaming/ crying	Learn 5 common signs of stress in dogs	Praise and treat puppy when they gently touch your hand	Play/feed puppy near or in very shallow water
Create platforms made of different textures for puppy to walk across	Let puppy see moving items (e.g. scooters, cars, children's toys)	Create a sensory area – fill pots with ornamental grasses and lavender	Change your appearance – wear a wig/ draw on a moustache or beard	Enjoy a quick game of tug	Blow bubbles for your puppy to chase	Hide treats in shallow boxes	Gently brush puppy and give them a chew as you groom
Let puppy interact with 10 different objects (e.g. saucepan, bottle)	Teach your puppy to wear their harness	Use one-third of puppy's daily food for training	Use food puzzle toys daily	Carry bulky items around the house (e.g. umbrella, cane, boxes)	Listen to sounds of traffic as puppy plays	Practise calmly having the lead put on/off	Listen to sounds of fireworks/ bangs at lowest volume. Over time, slowly increase volume
Say puppy's name and treat for eye contact	Teach puppy to be gentle when taking treats	Knock at the door/ring the doorbell and feed puppy treats on their bed	As puppy eats from the bowl, drop in a tastier treat	Let your puppy experience different types of weather	See/hear wheelie bins being moved	Rearrange furniture or add new items into the room	Let puppy explore under and over objects

LESSONS FOR FUTURE DISASTERS

The COVID-19 pandemic presented a new challenge for all of us, but there are many parallels and lessons to be learned from research into pet ownership during other disasters like hurricanes or bushfires. For example, commitment to animals lowers the odds of evacuation ahead of hurricanes, and puts owners at further risk when they try to rescue their pets.[14] Similar issues have been observed in Australian bushfires[15] and earthquakes in Japan.[16] In a US study, a strong attachment to a pet increased the likelihood of delaying seeking testing and treatment for COVID-19.[17] Fearing being unable to find accommodation for a pet, worrying if others will be able to care for the dog and the quality of care especially if pets had additional needs, increased the odds of a delay in seeking help even further.

In our survey, we found that only just over a third of the owners had some form of an emergency plan before the pandemic, including who would look after their dogs if needed (although over half of those who lived alone did). Even more worryingly, only 7 per cent more owners created an emergency plan in response to the pandemic. Such a plan is important: it can help to keep owners, pets and rescue services safe. The plan is recommended to include knowing how to evacuate the animals if needed, planning who will care for them if the owner can't, and having a Pet Emergency Survival Kit.[18] This kit should include 72 hours' worth of food, water and medicines, as well as blankets or towels, toys, lead, muzzle, pet carrier, medical records, vet's contact details, pet first aid kit, current photo of your dog with their name and address and important info (feeding schedule, temperament and behaviour, medical concerns, ID tags and microchip number) and a list of boarding facilities and hotels that can accommodate pets.

As owners, we need to plan ahead for what happens to our pets if something happens to us. Our pets are valued family members. We aren't willing to leave our dogs behind uncared for, and yet this could put our own health in danger. Neither are we prepared to suddenly stop walking them if advised to. Policymakers and governments need to bear these things in mind when creating rules and advice.

CHAPTER 15

TIME TO SAY GOODBYE

DEALING WITH AGEING AND LOSING YOUR PET

This is the hardest chapter to write, but I feel it is really important that we talk more openly about death, for the sake of ourselves and our dogs. Today's pet owners are more likely to treat their animals as members of the family rather than as workers who perform a specific task, which they might have done in the past. In fact, in my interviews with dog owners, some describe their dogs as like a child or grandchild. Others (particularly those who don't seem to like people much) argue that dogs are dogs. But to some of them, this means that they are even more important than people. I don't think we actually yet have a word to describe what a dog is to us, so comparing them to a family member will have to do for now.

While this emotional closeness to our animals is wonderful for most of their lives, it raises a difficult situation when it comes to them getting old, and ultimately their death.

WHEN IS A DOG OLD?

Dr Lisa Wallis is my post-doctoral researcher who has spent many years studying how dogs think and how they age, and more importantly, how we can help them age well. Ageing literally means getting older, but although it is a natural process, by the time our dogs go into their twilight years, they have accumulated changes and diseases which result in signs of failing limbs or organs. Lisa says:

> "My interviews with dog owners suggest that when we notice changes in our golden oldies, we mostly attribute them to the inevitable signs of 'getting old', and don't do a lot about it. However, the veterinary professionals I interview disagree that these changes are simply normal, believing there are underlying reasons for them, and while we may not be able to alleviate them all, there are many that can be addressed. As owners, it is up to us to notice their symptoms, and it is our responsibility to keep our old dogs in the best of health, so we can make their old age as comfortable and fulfilling as possible."

In humans, the term "senior" is just a definition of a person who has reached a certain age in their lifespan (usually 65 years). How old a dog will be when it becomes a "senior citizen", and its rate of ageing, depends on the dog's size, breed (pure or mixed/cross-breeds), sex, reproductive status (intact/neutered), lifestyle, body condition score and health status.[1] This means it can vary dramatically between individuals. In particular, large dogs die young because they age quickly.[2]

However, for simplicity, we can classify dogs according to their weight: dogs weighing under 23 kilograms are senior at around nine years of age, those between 23 and 40 kilograms at seven years, and dogs weighing over 40 kilograms at six years of age.[3]

However, "geriatric" is a term used more often in humans to refer to health status when a person is extremely fragile. Geriatric dogs are those that have entered the final 10 per cent of their expected lifespan. Since there are many parallels between old-age pets and elderly humans, we can use the same indication criteria to classify pets as geriatric, including

AGE IN YEARS	WEIGHT IN KIOGRAMS			
	<9KG	>9–23KG	>23–40KG	>40KG
5				
6				Senior
7			Senior	Senior
8			Senior	Geriatric
9	Senior	Senior		
10	Senior	Senior	Geriatric	
11	Senior		Geriatric	
12		Geriatric		
13		Geriatric		
14	Geriatric			
15	Geriatric			
16				
17				
18				
19				
20				

AGES AT WHICH DIFFERENT-SIZED DOGS BECOME CLASSED AS "SENIOR" AND "GERIATRIC"

a combination of decreased physical activity, slowed gait, fatigue, weakness, weight loss, social withdrawal, and increased vulnerability to physiological stresses. Aged dogs can also suffer from a type of doggy Alzheimer's or dementia, which is called canine cognitive dysfunction.[4]

HELPING OUR DOGS AGE GRACEFULLY

By the time bodily systems start to fail and clinical signs appear, it can often be too late to correct them. So, by anticipating the potential problems, and monitoring our dogs closely, especially during the earlier years when ageing effects are not so apparent, we can potentially increase the number of years of pleasure and companionship our dogs provide. Some changes can appear very suddenly, and others change very gradually and so are harder to detect. There are signs to look out for, which are listed below. There are also tools available to help you keep track of your dog's health, such as their response to medication and signs of pain and distress.[5]

Here are the most common physical and behavioural signs to watch out for as your dog gets older. This is based on our research with veterinary professionals and dog owners (currently unpublished), and other scientific knowledge.

Digestion – greater sensitivity to changes, increased/decreased appetite, preference for non-food items. A reduced metabolic rate means that they require less food to maintain weight, or need a diet with less energy (fewer calories). Increases/decreases in weight. Vomiting, salivation.

Muscle loss – weak, wobbly, prone to falls, pulled muscles/injuries.

Respiration – nasal discharge, cough/snoring/gagging, rasping/rapid laboured breathing/panting, frequent sneezing.

Defecation – straining, increased/decreased frequency/volume, bouts of diarrhoea, constipation, blood, mucus, changes in colour, faecal incontinence, discharge from anal glands, bleeding from anus.

Pain signs – trembling, pacing, licking joints, change in temperament (increased fear/aggression), depression/lethargy, decreased enthusiasm, increased panting, yelping, whining.

Coat – greying, thin, greasy, smelly, dry, bald patches, coarse.

Urination – straining, increased/decreased frequency/volume, blood in urine, urinary incontinence, discharge/blood coming from penis/vulva.

Impaired hearing – fails to respond when addressed, easily startled, may sleep more deeply. Look out for inflammation in ear, discharge, bad smell, itching, head shaking.

Skin – thickening/thinning, sores, calluses, lumps, swellings, warts, colour changes, dandruff.

Impaired mobility/pain signs – stiffness, limping, slowing down, uncomfortable, unwilling to walk/play, physical and muscle changes, postural/gait changes, sleeping more, difficulty to get in position to toilet, won't jump up on sofa/bed/car/boot.

Impaired eyesight – bumps into things, can't follow treat in hand. Look out for discharge from eyes, cloudiness, soreness, redness, warts, swelling, difference in size of pupils.

Paws – dry/cracked pads, redness, abscess between toes, dragging paw.

Dementia – confusion/disorientation, gets lost/stuck, less interested in petting, clingy/dependent, paces aimlessly, more anxious/irritable, awake at night, increases in vocal-izations, unresponsive/apathetic, loss of housetraining/unresponsive to previously known commands.

Dental problems – red gums, bad breath, tooth scale, discharge, growths, loose/cracked/discoloured teeth, facial swelling, reluctance to chew.

Nails – thickening/thinning, brittle, splitting, overly long due to reduced activity, uneven wear.

Our ability to recognize pain signs in dogs is of particular importance,[6] since 80 per cent of dogs aged eight or older will have signs of arthritis.[7] Lisa tells me, "My own dog Muffy, a Tibetan terrier, tore her cruciate ligament when she was six and had to have surgery. She later developed arthritis in that leg, and needed pain relief for the remainder of her life."

Elderly dogs and dogs with chronic health problems will benefit from seeing your vet more frequently than the once per year for vaccination and a check-up that well-cared-for younger dogs receive. Older dogs should be seen at least every six months. These check-ups give you the chance to bring your vet's attention to any minor changes you have noticed and to ask any questions you may have about your dog's health and care. Lisa suggests:

"Prior to your visit, write down any questions you would like to ask, keep a diary of health-related events, and use your phone to photograph or video any issues or behaviours that your dog may have to show your vet in your appointment. Always seek an appointment quickly if you notice anything out of the unusual that gives you concern, as this will give you the best chance to cure or control an emerging disease. Do not procrastinate – problems seldom go away on their own – an early diagnosis will not shorten your dog's life, or increase its suffering. When Muffy was 13 she became lethargic and started to get very fussy about her food. I knew something was wrong, as usually she had a very healthy appetite. After a quick trip to the vet, she was diagnosed with kidney disease, given a special diet and fluid therapy, and she was like a new dog, running and playing like a puppy! I am so glad that I got the extra time with her, which she wouldn't have had if I delayed taking her to the vet."

One of the best ways to prolong your dog's life is to keep it at a healthy weight, and on the "lean" side.[8] Excess fat places extra stresses on body systems, and many diseases of old age can be exacerbated by obesity. Older dogs do require less exercise as they age, but exercise is still of vital importance for maintaining muscle mass and range of movement of the limbs, and for environmental and mental enrichment (keeping the mind active and engaged). The type of exercise you might engage your dog in will depend on his level of capability and willingness. Lisa has more suggestions than the typical dog walk:

"Allowing an old dog to be in a different environment or providing enrichment for a few times a day, even if this is only going to be spending 10 minutes pottering and sniffing around a local lamppost or giving them their meal in a feeding toy rather than a bowl, will break up the monotony of the day. It will keep them alert, and give them something to look forward to, with the added bonus of increasing the dog-owner bond."

A study in laboratory dogs found that behavioural enrichment (which consisted of increased daily walking), environmental enrichment (dogs were housed with kennel mates and given different toys) and cognitive enrichment (the dogs repeatedly learnt to discriminate between two different objects) helped to preserve learning ability in aged dogs.[9] Therefore, similarly to people, enrichment can protect the working of our brains.[10]

The older dog may experience difficulties in the home when they start to suffer from reduced mobility or failing eyesight. The home environment can become increasingly hazardous, and present your ageing dog with significant challenges. These include steps and stairs, slippery floors, changes in furniture or room setups (especially after moving house), fires and heaters (old dogs often seek out warm places to rest and might overheat as they are less well able to maintain their body temperature), garden ponds (they might fall in), and people and other animals (especially younger dogs or children – older dogs might have a reduced tolerance level for disturbance by even the most well intentioned). There are things you can do to help your dog, which include:[11]

» using a ramp to get them in and out of the car;
» walking them proportionally more on-lead and planning rest stops;
» ensuring their toileting area is well-lit at night;
» giving them a dog bed stable and big enough to stretch out their whole body;

» providing non-slip flooring (like rubber-backed mats) around their bed and feeding areas;

» feeding them alone in a quiet place;

» raising their food and water bowls to the height of their shoulder so they don't need to lean forwards and lose their balance;

» placing secure rugs and runners on slippery floors;

» consider a stair-gate to stop them going up and down on their own;

» consider only allowing access to "safe" rooms when left alone, in particular to prevent unwanted jumping on sofas, beds and chairs if they shouldn't be doing this for medical reasons;

» modify play to low-impact games to avoid racing around or vigorous head shaking, such as using puzzle feeders.

Grooming your senior dog daily will allow you to examine your dog for signs of ill health, ensures contact between you and your dog, and helps to reduce soiling and odours. Many owners are unaware that their dog may suffer from dental disease, simply because they never look in their dog's mouth. Lisa admits, "I learnt the hard way with my own dog, as she had chronic dental disease, which may have contributed to her diagnosis of kidney failure in old age."[12] Your vet may need to scale and polish your dog's teeth, and then you can brush them as advised in Chapter 3. If your dog will not allow their teeth to be brushed, then oral sprays and mouthwashes (that can be added to the dog's water) are available as a compromise, but not nearly as effective. Also remember, when it comes to diet, a standard dry or wet food is unlikely to suit all older dogs, as older dogs have a reduced metabolic rate (they require less food to maintain weight), reduced digestive capacity (some proteins may be harder to digest) and the presence of disease means that different individuals will have different requirements to maximize the function of their particular organs. Always speak to your vet when considering whether to change your dog's diet.

RETIRING AN ASSISTANCE DOG

If your dog is trained to perform tasks that help you with your health, such as those we have discussed in this book, then deciding when to retire them from these duties can be difficult to navigate. Once again, it is imperative to put their health and well-being needs first, even if this means that you will have to learn how to cope without their help. Most assistance dogs retire from formal working life at around eight years old. Some people plan for this by training a second dog ready to take over – in some cases, the original dog is relieved and naturally lets them get on with it and in other cases the competition gives them a second wind. There may be some tasks that the dog still enjoys doing and can be allowed to continue, but other tasks it is best for them to stop earlier. Having an objective and trusted person such as your vet or trainer, to talk these issues through with, is important, and one of the reasons why I recommend having the support of an experienced organization when training an assistance dog.

THE IMPORTANCE OF
PLANNING FOR THE END

The overwhelming majority of pets are euthanized, a responsibility we don't often have resting upon us for our other family members. This adds complexity to our grief, but one that I take some comfort in, as at least we have some control over our dogs' suffering. I expect many readers will already have tears in their eyes reading this and remembering an old canine friend. Imagine if this animal was not just a pet, but one trained to help you for health reasons. Losing a dog that you literally can't live well without would make things even more difficult when they die. For those who also live alone with their pet, losing their animal can have an even deeper effect.

If we are talking about fostering a mutually beneficial synergistic relationship with our dogs, then this has to apply to end-of-life care too. If we're looking to our dogs to make our lives better, we owe it to them to make the right decisions on their behalf when they get too old or too ill to live with dignity. Only you can make the decision as to whether that time has come, and this will depend on the individual personality of the animal. I have a habit of thinking through "what ifs", and there are certainly dogs I have owned that I have considered keeping alive for longer, or working with them to adjust for a specific disability. But there are also dogs who I know would not have coped well were something major to happen.

With many of my dogs that have passed, it was clear to me when the time had come. The day my collie Ben woke up struggling to breathe because of a tumour in his throat was one such moment. He had finally been diagnosed six weeks earlier. I had had suspicions for some time, due to a strange cough he sometimes did, but nothing could be found when vets felt and looked down his throat. Eventually, he began making a strange gravelly noise when he was eating, and I took him in for a check-up and Marisol (whom we met in Chapter 3) finally felt the lump deep in his throat. I was grateful for the few weeks I got to enjoy with him knowing it was near the end, but very sad that he would never get to meet the tiny baby that I had just begun to feel moving inside me. Ben was a huge softie who would have loved a baby to stare at all day.

Sometimes, it isn't so clear cut, as was the case with Jasmyn. Deteriorating kidneys, tremors, toileting accidents – it all slowly added up over time. The crunch came when I decided that the neurological episodes that she began to experience had become distressing for her, as she paced, cried, walked into walls and fell over. They were probably also happening when I wasn't there to hold her still and comfort her.

Although it is hard to think about, it is important to consider these issues in advance. There are "Quality of Life" surveys available that can help you monitor the well-being of your pet as he ages and decide whether it is time.[13] When

A SPECIAL MOMENT IN JAS'S AND MY FINAL DAYS TOGETHER

deliberating over Jas, I was given a great piece of advice: "You won't regret doing it too soon, but you will be distressed if you leave it too late." Because of this, I was able to avoid the rushed, traumatic trip to the veterinary surgery. I was able to arrange a calm, quiet passing at home for Jasmyn, on her dog bed in the living room, eating freshly baked sausage rolls and chocolate (don't judge me – I know chocolate is poisonous to dogs, but she still tried to steal it and at this point, it really didn't matter). I don't think she even realized what was going on, and she was happy.

There will also be decisions to be made about what you want to do with the body, and if you have not thought about this (or discussed it with the other family members) in advance you could feel rushed and under pressure to make a decision at an upsetting time. Typically, animals are sent for joint cremation if no other instructions are received, but other options include bringing the body home, or individual cremation so that you can scatter the ashes, or keep them in your living room. Ben and Jas are under a special tree in our local woods, so that we can remember them when we walk there with their pals. I must admit her box stayed in our dining room for a few months before I was ready to take her there.

COPING WITH GRIEF

If you are struggling after the loss of a pet, that's OK. The importance of pets in our lives is becoming increasingly accepted and understood, to the point where some employers are introducing formal compassionate leave when someone loses a pet. Other pet owners often understand your significant loss, although some won't, as they may have had a different relationship with their animals. The stages of grief that can occur when we lose an important person can also apply when we lose our pets (denial, anger, bargaining, depression, acceptance), often with the added burden of immense guilt that we had to make the choice, and possibly because of lack of finances to treat the animal's condition. Feelings related to losing a pet may not only apply to death, but also when the pet has to be given up for adoption, or goes missing. There are many resources available to help, and even a dedicated pet bereavement support hotline and email counselling service in the UK.[14] Do talk through your feelings with someone, and don't feel alone.

It is also common for owners to become overwhelmed with grief when their dog hasn't even died yet. Anticipatory grief is the name given to the tumultuous set of feelings and reactions that occur when someone is expecting the death of a loved one. These emotions can be just as intense as the grief felt after a death. The most important thing to remember is that anticipatory grief is a normal process, even if it's not discussed as often as regular grief, and again there are resources that can help.[15]

Telling children that a pet has died can be particularly difficult. When Brandon was young, his Nanny's dog Opie died suddenly overnight. We found that well-intended stories about her having "gone on a star now" just confused him. Children can understand more of what is going on than expected, and I recommend always being truthful. When Jasmyn died around a year later, I bought a lovely children's story to help us talk through his memories of her.[16]

For some (including me), getting a new dog helps with the healing. One canine pal never replaces another, but they can distract from the feeling of bereavement and help create fun new memories. I think it also helps me that I typically have multiple dogs at a time. When I am sad, I still have a warm furry body to cuddle, and I can channel my energy into activities with the dogs I still have. Thus, one option is succession planning your acquisition of dogs so that you (or they) are never suddenly left without canine company.

CHAPTER 16

CONCLUSION

The idea that our connections with animals are impactful has gained particular traction in the veterinary field, and is now spilling over to human medical colleagues and wider society. A term you may have heard of, which has been bandied around a lot in the academic literature and public health campaigning in recent years, is "One Health". So much so that, in a recent university restructuring, my department changed its name to "Livestock and One Health". Typically used to illustrate how important animal health and welfare is for preventing zoonotic diseases (those that transmit between animals and people, such as causes of food poisoning), the term also increasingly includes recognizing the impact of our pets on our well-being.[1] I fully welcome this more integrated approach, in particular the importance laid on both human health and animal welfare. As I have argued throughout this book, to truly enhance one, we need the other.

It's surprising to me that until relatively recently in academic timelines, scientists (and funders) had not thought much about the effects of something so obvious, right there in our homes. Perhaps it wasn't a topic deemed objectively observable for scientific scrutiny – put simply, our pets are so close to us that it is hard to step back with enough detachment to study them. The pioneers in the field also usually did their research as an unfunded "pet" side project (excuse the pun) alongside their main research. However, at an American National Institutes of Health workshop as far back as 1988, presenters Alan Beck and Lawrence Glickman concluded, "No future study of human health should be considered comprehensive if the animals with which they share their lives are not included."[2] Yet, over 30 years later, I have lost count of the major human health datasets I have enquired about only to be told, "No, sorry, we didn't ask about their pets." For someone who regularly observes major positive and negative impacts on well-being caused by pets, it beggars belief. On the rare occasion that health researchers do ask about pets, it is lucky if there are different species recorded, let alone any measure of the strength or type of relationship the person has with the pet. Both aspects turn out to be critical to measure. Just because you live with a pet, it doesn't mean you interact with it a lot; a fish in a bowl is likely to affect your health quite differently to a dog. We still have a lot more to learn about exactly how and why our beloved pets impact our health, but the future of this research area looks bright and we are rapidly building a much fuller picture.

SETTING YOURSELF UP FOR SUCCESS

Although I can't teach you everything you need to know about sourcing, owning and training your dog in one book, hopefully you now have a good idea how to progress. If you are planning to get a dog, you should now have a far better chance of things working out well and of avoiding potential issues. If you were thinking about getting a dog but have since decided that you don't have the capacity and right time now for what is required, then this book has also met its purpose. Your time will come, with thought and intention, and likely with a successful outcome. In the meantime, leverage your "pet effect" instead by helping others with their dogs.

If you already have a dog, you and your dog should both know each other better as a result of reading this book, and you should now be able to see things a little more from your dog's point of view. Being with your dog should be fun for the majority of the time – this is the secret to being a happy dog owner. Enjoy having him around, even if it isn't always perfect. Owning a dog is a partnership and training is a two-way communication. Remember that you are both on the same side and so battles should never be necessary.

Whatever you decide to do with your dog, always think and question. If being with your dog is stressful and he is not acting how you would like, ask yourself why, and decide what you can do to avoid it happening again. If someone else recommends anything for your dog, or if you read something in a book or see it on television, never simply accept it. Consistently ask yourself, "Is this right for my dog?", "Will she learn the right things from it?", or "Will it strengthen our relationship?"

A WELL-TRAINED DOG OR
A WELL-BEHAVED DOG?

What is the difference? If you are waiting at a bus stop, for instance, a well-trained dog would lie down and stay when you asked him; a well-behaved dog would simply settle down of his own accord. A well-trained dog is happy to do the right thing because you ask him. A well-behaved dog is happy to do the right thing without being asked. Of course, a dog can be well trained as well as well behaved, but the latter is much easier for you and should be your ultimate goal as a dog owner, because it doesn't require you to constantly tell them what to do. The more you practise with your dog, teaching him how to behave in as many different situations as possible, the better behaved he will be, making your life easier

In order for him to be well behaved, you must ensure that you reward him at the right time for the right things. Every time he settles down in the house of his own accord, tell him he is good and give him a calm fuss. When he doesn't jump up at a person, give him a treat. Praise him for being close to you on a walk. Find situations where your dog naturally does the right thing and reward them. Rewards mean that your dog will repeat this behaviour more often in the future, and this is how you end up with a well-behaved as well as well-trained dog.

THE THREE-TO-ONE RULE

For every negative experience, try to ensure that your dog has three pleasant ones soon afterwards. For example:

» Each time you tell your dog "no", praise him when he is doing the right thing at least three times.
» If you are cross with your dog, have three short play sessions soon afterwards.
» If you accidentally stand on your dog's toe, call him to you three times for cuddles.

» If your dog has a bad experience with another dog, find three nice dogs for him to meet as soon as possible afterwards.

» If you interrupt your dog when he misbehaves, reward him three times for doing the right thing.

» If you tell your dog not to jump up, make sure that you fuss him at least three separate times for not jumping up.

It is easy in theory, but can be hard to remember to put it into practice. It means that your dog will learn what you are wanting because he is having more good experiences than bad. Your dog learns that he gets more attention for being good than for misbehaving. If he has a bad experience, he must have more good experiences to balance it. Just as important is if you have a stressful moment with him, you also need enjoyable moments to make up for it, so that on balance, you are a happy dog owner.

CAUTIOUS OPTIMISM

In order to achieve the health benefits that are possible with dog ownership, it has to be a mutually beneficial relationship that is constantly worked at. Yet the assumption that pets are always great for our health prevails, and can be quite dangerous for both human well-being and animal welfare, as this book has laid out. Anthrozoologist Professor Hal Herzog writes a popular blog for *Psychology Today*,[3] in which he regularly unpicks these claims and highlights studies that go against the believed status quo. When I asked Hal why there appears to be a media bias and popular opinion that "The Pet Effect" is proven beyond doubt, he said, "There are a few reasons. Human thinking is slanted by the information we are most frequently exposed to. Psychologists call this the 'availability bias'. People also like 'feel-good' stories about the healing powers of animals. As a result, the press gives lots of attention to studies that find, for example, that getting a dog will make you less depressed, and ignore the studies that don't show this."[4]

The other big problem is that researchers often don't publish studies that produce disappointing results. Hal highlights that "inconvenient findings are often relegated to the proverbial 'file drawer', where they never see the light of day, either intentionally or because it can be harder to get the study published." This is called "positive publication bias", and is a problem in all scientific fields. In 2013, Dr Maggie O'Haire at Purdue University reported that 100 per cent of published studies of the impact of the animal-assisted therapy on autistic children reported positive effects; this is a statistical impossibility, especially given many methodological weaknesses in those study designs.[5] Hal tells me, "It is critical that investigators publish all their results, whether positive, no difference or even negative – like you always do – and fortunately, this is now happening more."

What appears to be clear is that it is not a done deal; just because you own a dog it doesn't mean that you are automatically going to be happier and healthier. What matters is what kind of relationship you have with her, how your dog behaves, and whether you regularly walk with her. In this book, we have investigated situations where having a dog was impacting the owner's mental health in a beneficial way, and also those where a dog was making an owner's life stressful and being detrimental to their mental well-being. We have discussed reasons why dogs can encourage people to walk more, but also reasons why some dogs make their owners not want to walk them. In all honesty, no dog is perfect, and there will be both great times and not-so-great times with yours. The good news is, we have identified what we can practically do to encourage more positive than negative experiences during our dog-ownership journey.

BEING INTENTIONAL ABOUT DOG OWNERSHIP

There is a saying that money can't buy happiness. There is also a saying: "Whoever said money can't buy happiness forgot about puppies." The scientific evidence points to the first statement being true – at least up to the point where our basic needs are met comfortably, more money doesn't necessarily buy us more happiness.[6] What gives us happiness is what we do with the resource that we have. Likewise, with dog ownership, simply buying a dog isn't guaranteed to make you happy, as the evidence presented in this book suggests. What we need to do is make intentional and thoughtful choices: about where we get the dog from, what type of dog it is, how we spend our time with our dogs, and how and what we train them to do. Through these choices, we can build the right kind of relationship that will benefit our own health and well-being.

It's not for the faint-hearted. It's a lot of work. But you will get back what you put in, 10 times over. I truly believe that owning my dogs does make me happier. When I am on my deathbed I will think of all the dogs I shared my life with, and imagine the taste of sausage rolls and chocolate.

I can't wait to hear how this book has helped you to be a happier dog owner.

ACKNOWLEDGEMENTS

They say it takes a village, and this book is certainly a collection of expertise, experiences and research evidence created by very many people. Many thanks to all those who have taken the time to share their thoughts with me over the years, whether research participants, academic colleagues, professional colleagues, or friends and family. Of course, I am also extremely grateful to all those named people throughout the book who allowed me to share their personal and often vulnerable stories to illustrate important points, or even provided me with a quote to help explain a principle. These case histories and real conversations really helped the book to come alive.

I am extremely indebted to Erica Peachey, for taking me under her wing all those years ago and allowing me to observe, and then lead, many training classes and behavioural consultations in her practice. Thanks for all the long, philosophical musings over all things dogs and their people, especially on long car journeys which always fly by with you. Mostly, thanks for freely allowing me to plagiarize lots of your advice to dog owners from handouts and workbooks. Your dedication to increasing people's skills with dogs, over any thoughts of competition and personal success, has always been inspiring.

It may not feel like it but this book actually covers a broad range of subjects, of which I am not an expert in all. So many people contributed to the writing of different chapters, with their thorough and honest comments, criticisms, suggestions and praise. Special mentions go to Professor Daniel Mills, Dr Sara Owczarczak-Garstecka, Dr Taryn Graham, Dr Lisa Wallis, Selina Gibsone, Georgiana Woods, Karen Ingram, Dr Marisol Collins, Dr Claire Guest, Peter Gorbing, Professor Ryan Rhodes and Danielle Beck. Hopefully between us we didn't forget anything really important.

I am also very grateful to all the assistance dog trainers, handlers and dogs who have taught me so much in order to be able to write a book about it. Special thanks must be given for the support of Dr Fiona Cooke and Cass Peters. Thanks also to the members of the Association of Pet Behaviour Counsellors for contributing thoughts and ideas, including Katie Patmore for instructions on how to teach a dog to settle, Hanne Grice and Rachel Hill for resources on socialization tasks, Claire Stewart for creating guidance on finding a responsible breeder, and Karen Wild and Lisa Benn for supplying pictures to base illustrations on. I am also proud to be a founding member of the Merseyside Dog Safety Partnership who provide me with excellent insight and analysis of dog-related issues through our interdisciplinary forum.

There are also some more people not previously mentioned who I would like to thank for their particular support throughout my academic career leading to this book. These include my University of Liverpool colleagues including: my PhD supervisor and long-time collaborator Professor Rob Christley; Prof Alex German; Prof Nicola Williams; Dr Gina Pinchbeck; Prof Francine Watkins; Prof Liz Perkins; and Dr John Tulloch. To those outside my academic institution I am grateful for mentorship and collegiality from Dr Hayley Christian, Prof

Hal Herzog, Dr Maggie O'Haire, Prof Sandra McCune, Prof Nancy Gee, Prof James Serpell, Stephen Jenkinson, and Dr Layla Esposito. In addition, I must thank another of my PhD supervisors, Dr John Bradshaw, for sharing his book agent and opening the door to this crazy book idea becoming a reality.

Speaking of agents, Doug Young at PEW Literary has championed the idea of this book from the outset and kept me straight through wobbles of self-doubt. He "got it" when even I wasn't sure what this book was. In addition to Doug's insightful comments and pulling me back from "being too academic – too many numbers", thanks to my editor Issy Wilkinson at Welbeck for challenging me to deliver a clear message that is both evidence-based and constructive; to find the balance between my scientific pessimism and my optimism as dog owner and trainer, who believes in what dogs can do for people, under the right circumstances. Further acknowledgement goes to copy-editor Meredith Olson for helpful tightening, narrative flow and clarity. We actually managed to write a book in the middle of a pandemic! Many thanks also for the beautiful illustrations by Hannah George.

Finally my gratitude (and apologies) to my family for constantly mentioning so many of you and our dogs throughout this book, and perhaps some stories you would rather everyone forgot about. You, and our dogs, are clearly so important to me that I felt the need to tell the world all about us. We learned some lessons the hard way, so that hopefully others now don't have to.

FURTHER READING

Here is a list of my favourite books I can recommend if you want to find out more about the topics I have touched upon in this one.

Bradshaw, John (2011). *In Defence of Dogs: Why dogs need our understanding*. London: Penguin.
An exploration of the history and science of the dog–human relationship and argument for the use of kind, reward-based dog training methods, from a renowned anthrozoologist.

Todd, Zazie (2020). *WAG: The science of making your dog happy*. Vancouver: Greystone Books
Zazie is a terrific writer who uses her science and dog training knowledge to explain the latest evidence around dog welfare and training. She also writes a great blog at www.companion animalpsychology.com

Bradshaw, John (2017). *The Animals Among Us: The new science of anthrozoology*. London: Allen Lane
A deep examination of the history and scientific evidence around why we own pets and how they affect us.

Herzog, Hal (2010). *Some We Love, Some We Hate, Some We Eat: Why it's so hard to think straight about animals.* **London: HarperCollins**

My favourite non-fiction book that I read again and again. Hal is an engaging and funny anthrozoologist who takes us through decades of research and stories from himself and others about how we feel, think and act about animals in all sorts of contexts.

Merseyside Dog Safety Partnership; www.merseydogsafe.co.uk

The Merseyside Dog Safety Partnership is a group of experts in dog-bite prevention intent on helping communities keep their dogs, children, family and friends safe. I co-founded the group a few years ago and the website contains a number of resources on dog-bite prevention for the general public use (such as our advice sheets on "Help, I've been bitten by a dog" and "Help, my dog has bitten someone") as well as resources for professionals working in the sphere of dog-bite prevention (such as a "hazard perception" test video on safety around dogs, starring Roxie and Jasmyn). We also list our University of Liverpool dog-bite research papers here.

Mills, Daniel and Westgarth, Carri (2017). *Dog Bites: A multidisciplinary perspective.* **Sheffield: 5M Publishing**

This is a textbook about multiple viewpoints on the topics of dog bites, that we collated as topic chapters from over 30 different experts from many different fields. It includes dog behaviour and training knowledge but also public health, epidemiology, forensic science, psychology, sociology, genetics, law and medicine.

Pryor, Karen (2002). *Don't Shoot The Dog: The new art of teaching and training.* Letchworth: Ringpress Books
An old one but a very good one. Not actually much about dog training per se, but how to use learning theory and reinforcement in your interactions with people and animals, and even on yourself, to get the behaviour you desire.

Wynne, Clive (2019). *Dog is Love: The science of why and how your dog loves you.* London: Quercus
A fascinating perspective from a behavioural scientist who has spent many years studying dogs and their owners, about how dogs think and what they do, and why they are so successful; their capacity to love us, and get us to love them.

Daley-Olmert, Meg (2010). *Made for Each Other: The Biology of the Human–Animal bond.* Boston: Da Capo Press
A nice overview of human–animal bond research, in particular relating to the role of the hormone oxytocin. However, a critical perspective on the research is lacking somewhat.

Serpell, James (2016). *The Domestic Dog: Its evolution, behavior and interactions with people.* Cambridge: Cambridge University Press
Something of a bible for canine researchers and trainers, and now in its second edition.

Donaldson, Jean (1996). *The Culture Clash.* Berkeley: James and Kenneth Publishers
Now a fairly old book and written from a North American perspective at times where things can be a little different to the UK, however a great resource on training skills, why

dogs do what they do and how you should approach working with them.

Dunbar, Ian (2001). *BEFORE You Get Your Puppy*; and *AFTER You Get Your Puppy*. Berkeley: James and Kenneth Publishers
Free eBooks available at https://www.dogstardaily.com/training/you-get-your-puppy; https://www.dogstardaily.com/storefront/after-you-get-your-puppy.

A great guide to all the things you need to know and plan for when thinking about getting a puppy, and a second guide on what to do with it when you get it.

NOTES

CHAPTER 1

1 Fox, R. and N. R. Gee (2019). "Great expectations: changing social, spatial and emotional understandings of the companion animal–human relationship." *Social & Cultural Geography* **20**(1): 43–63. https://doi.org/10. 1080/14649365.2017.1347954

2 For example, in the UK: owners can be sent to prison if their dog kills someone (Dangerous Dogs Act 1991); it is an offence to walk your dog off-lead on a road (Road Traffic Act 1998); and compulsory for dogs to be microchipped and to wear a collar and tag engraved with the name and address of the owner (Control of Dogs Order 1992 and The Microchipping of Dogs England Regulations 2015)

CHAPTER 2

1 Bradshaw, J. (2018) *The Animals Among Us: The New Science of Anthrozoology*. London: Penguin

2 Pet Food Manufacturers Association (2019). Dog Population 2019. https://www.pfma.org.uk/dog-population-2019 (Accessed 26 Jan 2021)

3 American Veterinary Medical Association (2018) US Pet Ownership Statistics. https://www.avma.org/resources-tools/reports-statistics/us-pet-ownership-statistics (Accessed 26 Jan 2021)

4 Animal Medicines Australia (2019) "Pets in Australia: A national survey of pets and people." https://animalmedicinesaustralia.org.au/report/pets-in-australia-a-national-survey-of-pets-and-people (Accessed 26 Jan 2021)

5 Christian, H., C. Westgarth, A. Bauman, E. A. Richards, R. Rhodes, K. Evenson, J. A. Mayer and R. J. Thorpe (2013). "Dog ownership and physical activity: A review of the evidence." Journal of Physical Activity and Health **10**(5): 750–759. https://doi.org/10.1123/jpah.10.5.750

6 McNicholas, J., A. Gilbey, A. Rennie, S. Ahmedzai, J.-A. Dono and E. Ormerod (2005). "Pet ownership and human health: a brief review of evidence and issues." BMJ **331**(7527): 1252–1254. https://doi.org/10.1136/bmj.331.7527.1252

7 Dolan, P. and L. Kudrna (2016). "Sentimental Hedonism: Pleasure, Purpose, and Public Policy", in Vittersø J. (ed) *Handbook of Eudaimonic Well-Being*. International Handbooks of Quality-of-Life/Springer. Cham
For a deeper exploration of his views on happiness, read Dolan, P. (2014). *Happiness by Design: Finding Pleasure and Purpose in Everyday Life*. London: Penguin

8 Westgarth, C., R. M. Christley, G. Marvin and E. Perkins (2020). "Functional and recreational dog walking practices in the UK." Health Promotion International. https://doi.org/10.1093/heapro/daaa051

9 Daley Olmert, M. (2010). *Made for Each Other: The Biology of the Human–Animal Bond.* Da Capo Press; Beetz, A., K. Uvnäs-Moberg, H. Julius and K. Kotrschal (2012). "Psychosocial and Psychophysiological Effects of Human–Animal Interactions: The Possible Role of Oxytocin." *Frontiers in Psychology* 3: 234. https://doi.org/10.3389/fpsyg.2012.00234

10 Nagasawa, M., S. Mitsui, S. En, N. Ohtani, M. Ohta, Y. Sakuma, T. Onaka, K. Mogi and T. Kikusui (2015). "Oxytocin-gaze positive loop and the coevolution of human–dog bonds." *Science* 348(6232): 333–336. https://doi.org/ 10.1126/science.1261022

11 Vormbrock, J. K. and J. M. Grossberg (1988). "Cardiovascular effects of human-pet dog interactions." *Journal of Behavioral Medicine* 11(5): 509–517. https://doi.org/10.1007/BF00844843

12 Friedmann, E., A. H. Katcher, J. J. Lynch and S. A. Thomas (1980). "Animal companions and one-year survival of patients after discharge from a coronary-care unit." *Public Health Reports* 95(4): 307–312. https://www.jstor.org/stable/4596316

13 Serpell, J. (1991). "Beneficial effects of pet ownership on some aspects of human health and behaviour." *Journal of the Royal Society of Medicine* 84(12): 717–720. https://doi.org/10.1177/014107689108401208

14 One study estimated that for the year 2000, savings of €5.59 billion for Germany and $3.86 billion (AUD) for Australia were being made.
Headey, B., M. Grabka, J. Kelley, P. Reddy and Y. Tseng (2002). "Pet ownership is good for your health and saves public expenditure too: Australian and German longitudinal evidence." *Australian Social Monitor* 4: 93–99. https://www.semanticscholar.org/paper/Pet-Ownership-is-Good-for-Your-Health-and-Saves-and-Headey-Grabka/c943094b06aa1568bbf40501a6a4ab530c75801d

15 Hall, S., L. Dolling, K. Bristow, T. Fuller and D. S. Mills (2016). *Companion Animal Economics: The Economic Impact of Companion Animals in the UK.* CABI

16 Murray, J. K., W. J. Browne, M. A. Roberts, A. Whitmarsh and T. J. Gruffydd-Jones (2010). "Number and ownership profiles of cats and dogs in the UK." *Veterinary Record* 166(6): 163–168. https://bvajournals.onlinelibrary.wiley.com/doi/abs /10.1136/vr.b4712
Westgarth, C., G. L. Pinchbeck, J. W. S. Bradshaw, S. Dawson, R. M. Gaskell and R. M. Christley (2007). "Factors associated with dog ownership and contact with dogs in a UK community." *BMC Veterinary Research* 3(5). https://doi.org/10.1186/1746-6148-3-5

17 Saunders, J., L. Parast, S. H. Babey and J. V. Miles (2017). "Exploring the differences between pet and non-pet owners: Implications for human-animal interaction research and policy." *PLOS ONE* 12(6): e0179494. https://doi.org/10.1371/journal.pone.0179494

18 As well as previous references, see also: Westgarth, C., J. Heron, A. R. Ness, P. Bundred, R. M. Gaskell, K. P. Coyne, A. J. German, S. McCune and S. Dawson (2010). "Pet ownership during childhood: findings from a UK birth cohort and implications for public health research." *International Journal of Environmental Research and Public Health* 7(10): 3704–3729. https://

doi.org/10.3390/ijerph7103704; Westgarth, C., L. Boddy, G. Stratton, A. German, R. Gaskell, K. Coyne, P. Bundred, S. McCune and S. Dawson (2013). "Pet ownership, dog types and attachment to pets in 9–10 year old children in Liverpool, UK." *BMC Veterinary Research* 9(1): 102. https://doi.org/10.1186/1746-6148-9-102

19 Bao, K. J. and G. Schreer (2016). "Pets and Happiness: Examining the Association between Pet Ownership and Wellbeing." *Anthrozoös* 29(2): 283–296. https://doi.org/10.1080/08927936.2016.1152721

20 Miles, J. N. V., L. Parast, S. H. Babey, B. A. Griffin and J. M. Saunders (2017). "A Propensity-Score-Weighted Population-Based Study of the Health Benefits of Dogs and Cats for Children." *Anthrozoös* 30(3): 429–440. https://doi.org/10.1080/08927936.2017.1335103

21 Westgarth, C., J. Heron, A. R. Ness, P. Bundred, R. M. Gaskell, K. Coyne, A. J. German, S. McCune and S. Dawson (2012). "Is Childhood Obesity Influenced by Dog Ownership? No Cross-Sectional or Longitudinal Evidence." *Obesity Facts* 5(6): 833–844. https://doi.org/10.1159/000345963

22 Fall, T., R. Kuja-Halkola, K. Dobney, C. Westgarth and P. K. E. Magnusson (2019). "Evidence of large genetic influences on dog ownership in the Swedish Twin Registry has implications for understanding domestication and health associations." *Scientific Reports* 9(1): 7554. https://doi.org/10.1038/s41598-019-44083-9

23 There is a nice summary by writer Hal Herzog of the research here: Herzog, H. (2019). "Do Our Genes Really Affect Our Relationships With Pets?" *Psychology Today.* https://www.psychologytoday.com/gb/blog/animals-and-us/201905/do-our-genes-really-affect-our-relationships-pets (Accessed 26 Jan 2021)

CHAPTER 3

1 Mellor, D. J. and N. J. Beausoleil (2015). "Extending the 'Five Domains' model for animal welfare assessment to incorporate positive welfare states." *Animal Welfare* 24(3): 241–253. https://doi.org/10.7120/09627286.24.3.241
Mellor, D. J. (2016). "Updating Animal Welfare Thinking: Moving beyond the 'Five Freedoms' towards 'A Life Worth Living'." *Animals: an open access journal from MDPI* 6(3): 21. https://doi.org/10.3390/ani6030021

2 Cobb, M. L., A. Lill and P. C. Bennett (2020). "Not all dogs are equal: perception of canine welfare varies with context." *Animal Welfare* 29(1): 27–35. https://doi.org/10.7120/09627286.29.1.027

3 Mellor, D. J., N. J. Beausoleil, K. E. Littlewood, A. N. McLean, P. D. McGreevy, B. Jones and C. Wilkins (2020). "The 2020 Five Domains Model: Including Human–Animal Interactions in Assessments of Animal Welfare." *Animals* 10(10): 1870. https://doi.org/10.3390/ani10101870

4 For more information on this particular issue, see resources from the Canine Arthritis Management organization. https://caninearthritis.co.uk (Accessed 27 Jan 2021)

5 A useful blog post on this topic is: Hay Veterinary Group. "Do I have to vaccinate every year – can't I just titre test?" https://hayvets.co.uk/i-

vaccinate-every-year-cant-i-just-titre-test/?fbclid=IwAR0B9g_5BlEUcuesOlE
aE4m40RJZHbABfAKxwGlGFTmhkh2Lu5oakq_qY78

6 Such as those given by the World Small Animal Veterinary Association:
Day, M. J. (2017). "Small animal vaccination: a practical guide for vets
in the UK." *In Practice* 39(3): 110–118. https://wsava.org/wp-content/
uploads/2020/01/Small-animal-vaccination-a-practical-guide-for-vets-in-
the-UK-Michael-Day.pdf (Accessed 27 Jan 2021)

Day, M. J., M. C. Horzinek, R. D. Schultz and R. A. Squires (2016). "WSAVA
Guidelines for the vaccination of dogs and cats." *Journal of Small Animal
Practice* 57(1): E1–E45. https://wsava.org/wp-content/uploads/2020/01/
WSAVA-Vaccination-Guidelines-2015.pdf (Accessed 27 Jan 2021)

7 Some breeds also carry a gene that can make them sensitive to certain drugs,
so research your breed.

8 Glickman, L. T., N. W. Glickman, G. E. Moore, E. M. Lund, G. C. Lantz, and
B. M. Pressler (2011). "Association between chronic azotemic kidney disease
and the severity of periodontal disease in dogs." *Preventive Veterinary Medicine*
99(2–4), 193–200. https://doi.org/10.1016/j.prevetmed.2011.01.011

9 Dodd, S., N. Cave, S. Abood, A.-K. Shoveller, J. Adolphe and A. Verbrugghe
(2020). "An observational study of pet feeding practices and how these have
changed between 2008 and 2018." *Veterinary Record* 186(19): 643. https://
bvajournals.onlinelibrary.wiley.com/doi/abs/10.1136/vr.105828

10 Axelsson, E., A. Ratnakumar, M.-L. Arendt, K. Maqbool, M. T. Webster,
M. Perloski, O. Liberg, J. M. Arnemo, Å. Hedhammar and K. Lindblad-Toh
(2013). "The genomic signature of dog domestication reveals adaptation to
a starch-rich diet." *Nature* 495(7441): 360–364. https://doi.org/10.1038/
nature11837

11 Schlesinger, D. P. and D. J. Joffe (2011). "Raw food diets in companion
animals: A critical review." *The Canadian Veterinary Journal – La Revue
vétérinaire canadienne* 52(1): 50–54. https://www.ncbi.nlm.nih.gov/pmc/
articles/PMC3003575

12 Lenz, J., D. Joffe, M. Kauffman, Y. Zhang and J. LeJeune (2009). "Perceptions,
practices, and consequences associated with foodborne pathogens and the
feeding of raw meat to dogs." *The Canadian Veterinary Journal – La Revue
vétérinaire canadienne* 50(6): 637–643. https://www.ncbi.nlm.nih.gov/pmc/
articles/PMC2684052

13 Raw preparations have been found to contain many parasites and bacteria
and in far higher amounts than in commercial cooked dog food. Scientific
studies do show evidence that dogs fed raw meat are more likely to carry (and
likely shed) antibiotic-resistant bacteria, which is an issue for both owner
health as well as the animal, given that transmission can occur between the
two. Therapy dogs fed raw meat diets have been shown to be 22 times more
likely to test positive for salmonella.

van Bree, F. P. J., G. C. A. M. Bokken, R. Mineur, F. Franssen, M. Opsteegh,
J. W. B. van der Giessen, L. J. A. Lipman and P. A. M. Overgaauw (2018).
"Zoonotic bacteria and parasites found in raw meat-based diets for cats and
dogs." *Veterinary Record* 182(2): 50. https://bvajournals.onlinelibrary.wiley.
com/doi/abs/10.1136/vr.104535

Nemser, S. M., T. Doran, M. Grabenstein, T. McConnell, T. McGrath, R. Pamboukian, A. C. Smith, M. Achen, G. Danzeisen, S. Kim, Y. Liu, S. Robeson, G. Rosario, K. McWilliams Wilson and R. Reimschuessel (2014). "Investigation of Listeria, Salmonella, and toxigenic *Escherichia coli* in various pet foods." *Foodborne Pathogens and Disease* 11(9): 706–709. https://doi.org/10.1089/fpd.2014.1748

Schmidt, V. M., G. L. Pinchbeck, T. Nuttall, N. McEwan, S. Dawson and N. J. Williams (2015). "Antimicrobial resistance risk factors and characterisation of faecal *E. coli* isolated from healthy Labrador retrievers in the United Kingdom." *Preventive Veterinary Medicine* 119(1): 31–40. https://doi.org/10.1016/j.prevetmed.2015.01.013

Wedley, A. L., S. Dawson, T. W. Maddox, K. P. Coyne, G. L. Pinchbeck, P. Clegg, T. Nuttall, M. Kirchner and N. J. Williams (2017). "Carriage of antimicrobial resistant *Escherichia coli* in dogs: Prevalence, associated risk factors and molecular characteristics." *Veterinary Microbiology* 199: 23–30. https://doi.org/10.1016/j.vetmic.2016.11.017

Lefebvre, S. L., R. Reid-Smith, P. Boerlin and J. S. Weese (2008). "Evaluation of the risks of shedding *Salmonellae* and other potential pathogens by therapy dogs fed raw diets in Ontario and Alberta." *Zoonoses Public Health* 55(8–10): 470–480. https://doi.org/10.1111/j.1863-2378.2008.01145.x

14 Such, Z. R. and A. J. German (2015). "Best in show but not best shape: a photographic assessment of show dog body condition." *Veterinary Record* 177(5): 125. https://bvajournals.onlinelibrary.wiley.com/doi/abs/10.1136/vr.103093

15 Salt, C., P. J. Morris, D. Wilson, E. M. Lund and A. J. German (2019). "Association between life span and body condition in neutered client-owned dogs." *Journal of Veterinary Internal Medicine* 33(1): 89–99. https://doi.org/10.1111/jvim.15367

16 A useful tool where you can enter your dog's breed (or closest) is available here: https://royalcanin.co.uk/body-condition-score.

17 Linder, D. E., L. M. Freeman, P. Morris, A. J. German, V. Biourge, C. Heinze and L. Alexander (2012). "Theoretical evaluation of risk for nutritional deficiency with caloric restriction in dogs." *Veterinary Quarterly* 32(3–4): 123–129. https://doi.org/10.1080/01652176.2012.733079

Linder, D. E., L. M. Freeman, S. L. Holden, V. Biourge and A. J. German (2013). "Status of selected nutrients in obese dogs undergoing caloric restriction." *BMC Veterinary Research* 9: 219. https://doi.org/10.1186/1746-6148-9-219

German, A. J., S. L. Holden, S. Serisier, Y. Queau and V. Biourge (2015). "Assessing the adequacy of essential nutrient intake in obese dogs undergoing energy restriction for weight loss: a cohort study." *BMC Veterinary Research* 11: 253. https://doi.org/10.1186/s12917-015-0570-y

18 WALTHAM Puppy Growth Charts. https://www.waltham.com/resources/puppy-growth-charts (Accessed 29 Jan 2021

Salt, C., P. J. Morris, R. F. Butterwick, E. M. Lund, T. J. Cole and A. J. German (2020). "Comparison of growth patterns in healthy dogs and dogs in abnormal body condition using growth standards." *PLOS ONE* 15(9): e0238521. https://doi.org/10.1371/journal.pone.0238521

19 Kersbergen, I., A. J. German, C. Westgarth and E. Robinson (2019). "Portion size and meal consumption in domesticated dogs: An experimental study." *Physiology & Behavior* 204: 174–179. https://doi.org/10.1016/j.physbeh.2019.02.034

20 Raffan, E., et al (2016). "A Deletion in the Canine POMC Gene Is Associated with Weight and Appetite in Obesity-Prone Labrador Retriever Dogs." *Cell Metabolism* 23(5): 893–900. https://doi.org/10.1016/j.cmet.2016.04.012

21 Shepherd, K. *The Canine Commandments*. https://www.kendalshepherd.com/books/the-canine-commandments (Accessed 29 Jan 2021)

22 Roulaux, P. E. M., I. R. van Herwijnen and B. Beerda (2020). "Self-reports of Dutch dog owners on received professional advice, their opinions on castration and behavioural reasons for castrating male dogs." *PLOS ONE* 15(6): e0234917. https://doi.org/10.1371/journal.pone.0234917

Farhoody, P., I. Mallawaarachchi, P. M. Tarwater, J. A. Serpell, D. L. Duffy and C. Zink (2018). "Aggression toward Familiar People, Strangers, and Conspecifics in Gonadectomized and Intact Dogs." *Frontiers in Veterinary Science* 5(18). https://doi.org/10.3389/fvets.2018.00018

23 Hart, B. L., L. A. Hart, A. P. Thigpen and N. H. Willits (2020). "Assisting Decision-Making on Age of Neutering for 35 Breeds of Dogs: Associated Joint Disorders, Cancers, and Urinary Incontinence." *Frontiers in Veterinary Science* 7(388). https://doi.org/10.3389/fvets.2020.00388

24 McGreevy, P. D., B. Wilson, M. J. Starling and J. A. Serpell (2018). "Behavioural risks in male dogs with minimal lifetime exposure to gonadal hormones may complicate population-control benefits of desexing." *PLOS ONE* 13(5): e0196284. https://doi.org/10.1371/journal.pone.0196284

25 Starling, M., A. Fawcett, B. Wilson, J. Serpell and P. McGreevy (2019). "Behavioural risks in female dogs with minimal lifetime exposure to gonadal hormones." *PLOS ONE* 14(12):e0223709. https://doi.org/10.1371/journal.pone.0223709

26 Packer, R. M. A., D. Murphy and M. J. Farnworth (2017). "Purchasing popular purebreds: investigating the influence of breed-type on the pre-purchase motivations and behaviour of dog owners." *Animal Welfare* 26(2): 191–201. https://doi.org/10.7120/09627286.26.2.191

27 Packer, R. M. A., D. G. O'Neill, F. Fletcher and M. J. Farnworth (2019). "Great expectations, inconvenient truths, and the paradoxes of the dog–owner relationship for owners of brachycephalic dogs." *PLOS ONE* 14(7): e0219918. https://doi.org/10.1371/journal.pone.0219918

28 Westgarth, C., R. M. Christley, G. Marvin and E. Perkins (2019). "The Responsible Dog Owner: The Construction of Responsibility." *Anthrozoös* 32(5): 631–646. https://doi.org/10.1080/08927936.2019.1645506

CHAPTER 4

1 Levine, G. N., K. Allen, L. T. Braun, H. E. Christian, E. Friedmann, K. A. Taubert, S. A. Thomas, D. L. Wells, R. A. Lange, C. Amer Heart Assoc and N. Council Cardiovasc Stroke (2013). "Pet Ownership and Cardiovascular Risk: A Scientific Statement From the American Heart Association." *Circulation* 127(23): 2353–2363. https://doi.org/10.1161/CIR.0b013e31829201e1

2 World Health Organisation (2020). https://www.who.int/news-room/fact-sheets/detail/physical-activity (Accessed 29 Jan 2021).

3 Scholes, S. and A. Neave. (2016). "Health Survey for England 2016: Physical Activity in Adults." https://files.digital.nhs.uk/publication/m/3/hse16-adult-phy-act.pdf (Accessed 21 Nov 2018)

4 Richards, E., P. Troped and E. Lim (2014). "Assessing the Intensity of Dog Walking and Impact on Overall Physical Activity: A Pilot Study Using Accelerometry." *Open Journal of Preventive Medicine* 4(7): 523–528. https://www.scirp.org/journal/paperinformation.aspx?paperid=47745

5 Westgarth, C., R. M. Christley, C. Jewell, A. J. German, L. M. Boddy and H. E. Christian (2019). "Dog owners are more likely to meet physical activity guidelines than people without a dog: An investigation of the association between dog ownership and physical activity levels in a UK community." *Scientific Reports* 9(1): 5704. https://doi.org/10.1038/s41598-019-41254-6

6 Cutt, H., B. Giles-Corti, M. Knuiman, A. Timperio and F. Bull (2008). "Understanding dog owners' increased levels of physical activity: Results from RESIDE." *American Journal of Public Health* 98(1): 66–69. https://doi.org/10.2105/AJPH.2006.103499

7 Sixty-five per cent of dog walkers achieved their 150 minutes through dog walking alone. Overall, 87 per cent of dog owners were found to meet physical activity guidelines, compared to only 63 per cent of the study participants who didn't own a dog.
Westgarth, C., R. M. Christley, C. Jewell, A. J. German, L. M. Boddy and H. E. Christian (2019). "Dog owners are more likely to meet physical activity guidelines than people without a dog: An investigation of the association between dog ownership and physical activity levels in a UK community." *Scientific Reports* 9(1): 5704. https://doi.org/10.1038/s41598-019-41254-6

8 Christian, H., C. Westgarth, A. Bauman, E. A. Richards, R. Rhodes, K. Evenson, J. A. Mayer and R. J. Thorpe (2013). "Dog ownership and physical activity: A review of the evidence." *Journal of Physical Activity and Health* 10(5): 750–759. https://doi.org/10.1123/jpah.10.5.750

9 Cutt, H. E., M. W. Knuiman and B. Giles-Corti (2008). "Does getting a dog increase recreational walking?" *International Journal of Behavioral Nutrition and Physical Activity* 5(1): 17. https://doi.org/10.1186/1479-5868-5-17

10 Allen, K., B. E. Shykoff and J. L. Izzo (2001). "Pet Ownership, but Not ACE Inhibitor Therapy, Blunts Home Blood Pressure Responses to Mental Stress." *Hypertension* 38(4): 815–820. https://doi.org/10.1161/hyp.38.4.815

11 Potter, K., J. E. Teng, B. Masteller, C. Rajala and L. B. Balzer (2019). "Examining How Dog 'Acquisition' Affects Physical Activity and Psychosocial Well-Being: Findings from the BuddyStudy Pilot Trial." *Animals* 9(9): 14. https://doi.org/10.3390/ani9090666

12 For examples, see: Thorpe, R. J., Jr, E. M. Simonsick, J. S. Brach, H. Ayonayon, S. Satterfield, T. B. Harris, M. Garcia, S. B. Kritchevsky and Health Aging and Body Composition Study (2006). "Dog ownership, walking behavior, and maintained mobility in late life." *Journal of the American Geriatrics Society* 54(9): 1419–1424. https://doi.org/10.1111/

j.1532-5415.2006.00856.x; Dall, P. M., S. L. H. Ellis, B. M. Ellis, P. M. Grant, A. Colyer, N. R. Gee, M. H. Granat and D. S. Mills (2017). "The influence of dog ownership on objective measures of free-living physical activity and sedentary behaviour in community-dwelling older adults: a longitudinal case-controlled study." *BMC Public Health* 17(1): 496. https://doi.org/10.1186/s12889-017-4422-5.

13 Salmon, J., A. Timperio, B. Chu and J. Veitch (2010). "Dog Ownership, Dog Walking, and Children's and Parents' Physical Activity." *Research Quarterly for Exercise and Sport* 81(3): 264–271. https://www.tandfonline.com/doi/abs/10.1080/02701367.2010.10599674

14 Christian, H., G. Trapp, K. Villanueva, S. R. Zubrick, R. Koekemoer and B. Giles-Corti (2014). "Dog walking is associated with more outdoor play and independent mobility for children." *Preventive Medicine* 67: 259–263. https://doi.org/10.1016/j.ypmed.2014.08.002

These findings were also echoed in my small sample of UK children (46), in which dog-owning children reported significantly more minutes per week walking (285 minutes) and free-time unstructured activity (e.g. playing, 260 minutes) than children without a dog:

Westgarth, C., R. M. Christley, C. Jewell, A. J. German, L. M. Boddy and H. E. Christian (2019). "Dog owners are more likely to meet physical activity guidelines than people without a dog: An investigation of the association between dog ownership and physical activity levels in a UK community." *Scientific Reports* 9(1): 5704. https://doi.org/10.1038/s41598-019-41254-6

15 See: Martin, K. E., L. Wood, H. Christian and G. S. A. Trapp (2015). "Not Just 'Walking the Dog': Dog Walking and Pet Play and Their Association With Recommended Physical Activity Among Adolescents." *American Journal of Health Promotion* 29(6): 353–356. https://doi.org/10.4278/ajhp.130522-ARB-262; Westgarth, C., A. R. Ness, C. Mattocks and R. M. Christley (2017). "A Birth Cohort Analysis to Study Dog Walking in Adolescence Shows No Relationship with Objectively Measured Physical Activity." *Frontiers in Veterinary Science* 4(62). https://doi.org/10.3389/fvets.2017.00062

16 Westgarth, C., J. Heron, A. R. Ness, P. Bundred, R. M. Gaskell, K. Coyne, A. J. German, S. McCune and S. Dawson (2012). "Is Childhood Obesity Influenced by Dog Ownership? No Cross-Sectional or Longitudinal Evidence." *Obesity Facts* 5(6): 833–844. https://doi.org/10.1159/000345963; Gadomski, A. M., M. B. Scribani, N. Krupa and P. Jenkins (2017). "Pet dogs and child physical activity: the role of child–dog attachment." *Pediatric Obesity* 12(5): E37–E40. https://doi.org/10.1111/ijpo.12156

17 Westgarth, C., L. M. Boddy, G. Stratton, A. J. German, R. M. Gaskell, K. P. Coyne, P. Bundred, S. McCune and S. Dawson (2017). "The association between dog ownership or dog walking and fitness or weight status in childhood." *Pediatric Obesity* 12(6). https://doi.org/10.1111/ijpo.12176

18 Coleman, K. J., D. E. Rosenberg, T. L. Conway, J. F. Sallis, B. E. Saelens, L. D. Frank and K. Cain (2008). "Physical activity, weight status, and neighborhood characteristics of dog walkers." *Preventive Medicine* 47(3): 309–312. https://doi.org/10.1016/j.ypmed.2008.05.007

19 Mubanga, M., L. Byberg, C. Nowak, A. Egenvall, P. K. Magnusson, E. Ingelsson and T. Fall (2017). "Dog ownership and the risk of cardiovascular disease and death – a nationwide cohort study." *Scientific Reports* 7(1): 15821. https://doi.org/10.1038/s41598-017-16118-6

20 Mubanga, M., L. Byberg, A. Egenvall, E. Ingelsson and T. Fall (2019). "Dog Ownership and Survival After a Major Cardiovascular Event: A Register-Based Prospective Study." *Circulation-Cardiovascular Quality and Outcomes* 12(10): 9. https://doi.org/10.1161/CIRCOUTCOMES.118.005342

21 It is also worth noting that researchers in these types of studies do not know who you are, and in fact we aren't interested in that, but in the patterns across the data as a whole. While understandably a controversial issue in terms of privacy fears, consenting to such "big data" collection and analysis, such as allowing your doctor's records to be used in such research, is critical if we want to make the kind of scientific advances that will improve population health.

22 Delicano, R. A., U. Hammar, A. Egenvall, C. Westgarth, M. Mubanga, L. Byberg, T. Fall and B. Kennedy (2020). "The shared risk of diabetes between dog and cat owners and their pets: register based cohort study." *BMJ* 371: m4337. https://doi.org/10.1136/bmj.m4337

23 Ding, D., A. E. Bauman, C. Sherrington, P. D. McGreevy, K. M. Edwards and E. Stamatakis (2018). "Dog Ownership and Mortality in England: A Pooled Analysis of Six Population-based Cohorts." *American Journal of Preventive Medicine* 54(2): 289-293. https://doi.org/10.1016/j.amepre.2017.09.012

24 Kramer, C. K., S. Mehmood and R. S. Suen (2019). "Dog Ownership and Survival: A Systematic Review and Meta-Analysis." *Circulation-Cardiovascular Quality and Outcomes* 12(10): 8. https://doi.org/10.1161/CIRCOUTCOMES.119.005554

25 Eller, E., S. Roll, C. M. Chen, O. Herbarth, H. E. Wichmann, A. von Berg, U. Kramer, M. Mommers, A. Thijs, A. Wijga, B. Brunekreef, M. P. Fantini, F. Bravi, F. Forastiere, D. Porta, J. Sunyer, M. Torrent, A. Host, S. Halken, K. C. L. Carlsen, K. H. Carlsen, M. Wickman, I. Kull, U. Wahn, S. N. Willich, S. Lau, T. Keil, J. Heinrich, G. A. Working Group and L. E. N. W. P.-B. Cohorts (2008). "Meta-analysis of determinants for pet ownership in 12 European birth cohorts on asthma and allergies: a GA(2)LEN initiative." *Allergy* 63(11): 1491-1498. https://doi.org/10.1111/j.1398-9995.2008.01790.x

26 Collin, S. M., R. Granell, C. Westgarth, J. Murray, E. S. Paul, J. A. C. Sterne and A. J. Henderson (2015). "Associations of Pet Ownership with Wheezing and Lung Function in Childhood: Findings from a UK Birth Cohort." *PLOS ONE* 10(6). https://doi.org/10.1371/journal.pone.0127756

27 Collin, S. M., R. Granell, C. Westgarth, J. Murray, E. Paul, J. A. C. Sterne and A. J. Henderson (2015). "Pet ownership is associated with increased risk of non-atopic asthma and reduced risk of atopy in childhood: findings from a UK birth cohort." *Clinical and Experimental Allergy* 45(1): 200-210. https://doi.org/10.1111/cea.12380

28 Chen, C.-M., C. Tischer, M. Schnappinger and J. Heinrich (2010). "The role of cats and dogs in asthma and allergy – a systematic review." *International Journal of Hygiene and Environmental Health* 213(1): 1-31. https://doi.org/10.1016/j.ijheh.2009.12.003

29 Westgarth, C., C. J. Porter, L. Nicolson, R. J. Birtles, N. J. William, C. A. Hart, G. L. Pinchbeck, R. M. Gaskell, R. M. Christley and S. Dawson (2009). "Risk factors for *Campylobacter upsaliensis* carriage in pet dogs in a community in Cheshire." *Veterinary Record* **165**: 526–530. https://bvajournals.onlinelibrary.wiley.com/doi/10.1136/vr.165.18.526

30 Tulloch, J. S. P., S. C. Owczarczak-Garstecka, K. M. Fleming, R. Vivancos and C. Westgarth (2021). "English hospital episode data analysis (1998–2018) reveal that the rise in dog bite hospital admissions is driven by adult cases." *Scientific Reports* **11**(1): 1767. https://doi.org/10.1038/s41598-021-81527-7

31 Westgarth, C., M. Brooke and R. M. Christley (2018). "How many people have been bitten by dogs? A cross-sectional survey of prevalence, incidence and factors associated with dog bites in a UK community." *Journal of Epidemiology and Community Health* **72**: 331–336. http://dx.doi.org/10.1136/jech-2017-209330

32 Westgarth, C. and F. Watkins (2017). "Chapter 23: Impact of dog aggression on victims", *Dog Bites: A Multidisciplinary Perspective*. Sheffield: 5M Publishing, pp. 309–320

33 In an online survey about dog-bite incidents, we found that at least 8 per cent of the dogs were euthanized, 4 per cent rehomed, and 2 per cent seized by police.
Oxley, J. A., R. Christley and C. Westgarth (2018). "Contexts and consequences of dog-bite incidents." *Journal of Veterinary Behavior: Clinical Applications and Research* **23**(Supplement C): 33–39. https://doi.org/10.1016/j.jveb.2017.10.005

34 Meints, K., C. Syrnyk and T. De Keuster (2010). "Why do children get bitten in the face?" *Injury Prevention* **16**(Supplement 1): A172–A173, http://dx.doi.org/10.1136/ip.2010.029215.617.

35 Meints, K., K. Allen and C. Watson (2010). "Atypical face-scan patterns in children misinterpreting dogs facial expressions evidence from eye-tracking." *Injury Prevention* **16**(Supplement 1): A173–A173. http://dx.doi.org/10.1136/ip.2010.029215.619

36 Herzog, H. (2018). "Tripping Over Pets Can Have Life-Changing Consequences." *Psychology Today*. https://www.psychologytoday.com/gb/blog/animals-and-us/201810/tripping-over-pets-can-have-life-changing-consequences (Accessed 30 Jan 2021)

37 Stevens, J. A., S. L. Teh and T. Haileyesus (2010). "Dogs and cats as environmental fall hazards." *Journal of Safety Research* **41**(1): 69–73. https://doi.org/10.1016/j.jsr.2010.01.001

38 A total of 54 cattle attacks were reported in the UK media from 1 January 1993 to 31 May 2013.
Fraser-Williams, A. P., K. M. McIntyre and C. Westgarth (2016). "Are cattle dangerous to walkers? A scoping review." *Injury Prevention* **22** (6). http://dx.doi.org/10.1136/injuryprev-2015-041784

CHAPTER 5

1 Powell, L., D. Chia, P. McGreevy, A. L. Podberscek, K. M. Edwards, B. Neilly, A. J. Guastella, V. Lee and E. Stamatakis (2018). "Expectations for dog ownership: Perceived physical, mental and psychosocial health consequences among prospective adopters." *PLOS ONE* **13**(7): 13. https://doi.org/10.1371/journal.pone.0200276

2 Westgarth, C., R. M. Christley, G. Marvin and E. Perkins (2017). "I Walk My Dog Because It Makes Me Happy: A Qualitative Study to Understand Why Dogs Motivate Walking and Improved Health." *Int J Environ Res Public Health* **14**(8). https://doi.org/10.3390/ijerph14080936

3 Allen, K., J. Blascovich and W. B. Mendes (2002). "Cardiovascular reactivity and the presence of pets, friends, and spouses: The truth about cats and dogs." *Psychosomatic Medicine* **64**(5): 727–739. https://journals.lww.com/psychosomaticmedicine/Abstract/2002/09000/Cardiovascular_Reactivity_and_the_Presence_of.5.aspx

4 For more poems by Matt Black, including his themed book on dogs, *Sniffing Lamp-posts By Moonlight*, go to https://www.matt-black.co.uk/. For a recorded performance of "The Snoopy Question: One Dog's Answer to World Peace" at the International Society for Anthrozoology Virtual Conference 2020, go to https://www.youtube.com/watch?v=Rp0zrpYg_i8&t=10s

5 Brooks, H., K. Rushton, S. Walker, K. Lovell and A. Rogers (2016). "Ontological security and connectivity provided by pets: a study in the self-management of the everyday lives of people diagnosed with a long-term mental health condition." *BMC Psychiatry* **16**(1): 409. https://doi.org/10.1186/s12888-016-1111-3

6 Barcelos, A. M., N. Kargas, J. Maltby, S. Hall and D. S. Mills (2020). "A framework for understanding how activities associated with dog ownership relate to human well-being." *Scientific Reports* **10**(1): 12. https://doi.org/10.1038/s41598-020-68446-9

7 Buller, K. and K. C. Ballantyne (2020). "Living with and loving a pet with behavioral problems: Pet owners' experiences." *Journal of Veterinary Behavior* **37**: 41–47. https://doi.org/10.1016/j.jveb.2020.04.003

8 Gonzalez-Ramirez, M. T., M. Vanegas-Farfano and R. Landero-Hernandez (2018). "Differences in stress and happiness between owners who perceive their dogs as well behaved or poorly behaved when they are left alone." *Journal of Veterinary Behavior: Clinical Applications and Research* **28**: 1–5. https://doi.org/10.1016/j.jveb.2018.07.010

9 Sundman, A.-S., E. Van Poucke, A.-C. Svensson Holm, Å. Faresjö, E. Theodorsson, P. Jensen and L. S. V. Roth (2019). "Long-term stress levels are synchronized in dogs and their owners." *Scientific Reports* **9**(1): 7391. https://doi.org/10.1038/s41598-019-43851-x

10 Beetz, A., K. Uvnäs-Moberg, H. Julius and K. Kotrschal (2012). "Psychosocial and psychophysiological effects of human–animal interactions: the possible role of oxytocin." *Frontiers in Psychology* **3**: 234. https://doi.org/10.3389/fpsyg.2012.00234

11 Fraser, G., Y. S. Huang, K. Robinson, M. S. Wilson, J. Bulbulia and C. G. Sibley (2020). "New Zealand Pet Owners' Demographic Characteristics, Personality,

and Health and Wellbeing: More Than Just a Fluff Piece." *Anthrozoös* 33(4): 561–578. https://doi.org/10.1080/08927936.2020.1771060

12 Enmarker, I., O. Hellzen, K. Ekker and A. G. T. Berg (2015). "Depression in older cat and dog owners: the Nord-Trøndelag Health Study (HUNT)-3." *Aging & Mental Health* 19(4): 347–352. https://doi.org/10.1080/136078 63.2014.933310; Parslow, R. A., A. F. Jorm, H. Christensen, B. Rodgers and P. Jacomb (2005). "Pet ownership and health in older adults: Findings from a survey of 2,551 community-based Australians aged 60–64." *Gerontology* 51(1): 40–47. https://doi.org/10.1159/000081433; Sharpley, C., N. Veronese, L. Smith, G. F. López-Sánchez, V. Bitsika, J. Demurtas, S. Celotto, V. Noventa, P. Soysal, A. T. Isik, I. Grabovac and S. E. Jackson (2020). "Pet ownership and symptoms of depression: A prospective study of older adults." *Journal of Affective Disorders* 264: 35–39. https://doi.org/10.1016/j.jad.2019.11.134

13 Mein, G. and R. Grant (2018). "A cross-sectional exploratory analysis between pet ownership, sleep, exercise, health and neighbourhood perceptions: the Whitehall II cohort study." *BMC Geriatrics* 18(1): 176. https://doi.org/10.1186/s12877-018-0867-3

14 Powell, L., K. Edwards, P. McGreevy, A. Bauman, A. Podberscek, B. Neilly, C. Sherrington and E. Stamatakis (2019). "Companion dog acquisition and mental well-being: a community-based three-arm controlled study." *BMC Public Health* 19(1). https://doi.org/10.1186/s12889-019-7770-5

15 Pereira, J. M. and D. Fonte (2018). "Pets enhance antidepressant pharmacotherapy effects in patients with treatment resistant major depressive disorder." *Journal of Psychiatric Research* 104: 108–113. https://doi.org/10.1016/j.jpsychires.2018.07.004

16 Evans-Wilday, A. S., S. S. Hall, T. E. Hogue and D. S. Mills (2018). "Self-disclosure with Dogs: Dog Owners' and Non-dog Owners' Willingness to Disclose Emotional Topics." *Anthrozoös* 31(3): 353–366. https://doi.org/10.1080/08927936.2018.1455467

17 Stanley, I. H., Y. Conwell, C. Bowen and K. A. Van Orden (2014). "Pet ownership may attenuate loneliness among older adult primary care patients who live alone." *Aging & Mental Health* 18(3): 394–399. https://doi.org/10.1080/13607863.2013.837147

18 Enmarker, I., O. Hellzen, K. Ekker and A. G. T. Berg (2015). "Depression in older cat and dog owners: the Nord-Trondelag Health Study (HUNT)-3." *Aging & Mental Health* 19(4): 347–352. https://doi.org/10.1080/13607863.2014.933310

In a small Australian study, the authors admit that once educational levels were adjusted for, the reduced levels of loneliness in new dog owners compared to control groups was no longer statistically significant.

Powell, L., K. Edwards, P. McGreevy, A. Bauman, A. Podberscek, B. Neilly, C. Sherrington and E. Stamatakis (2019). "Companion dog acquisition and mental well-being: a community-based three-arm controlled study." BMC Public Health 19(1). https://doi.org/10.1186/s12889-019-7770-5

19 Messent, P. R. (1983). "Social facilitation of contact with other people by pet dogs", in Katcher, A. H. and A. M. Beck, New Perspectives on our Lives

with Companion Animals. Philadelphia: University of Pennsylvania Press, pp.37–46; McNicholas, J. and G.M. Collis (2000) "Dogs as catalysts for social interactions: Robustness of the effect." *British Journal of Psychology* 91(1): 61–70. https://doi.org/10.1348/000712600161673

20 Wells, D. (2004). "The Facilitation of Social Interactions by Domestic Dogs." *Anthrozoös* 17: 340-352. https://doi.org/10.2752/089279304785643203

21 Antonacopoulos, N. M. D. and T. A. Pychyl (2014). "An Examination of the Possible Benefits for Well-Being Arising from the Social Interactions that Occur while Dog Walking." *Society & Animals* 22(5): 459–480. https://doi.org/10.1163/15685306-12341338

22 Wood, L., B. Giles-Corti and M. Bulsara (2005). "The pet connection: Pets as a conduit for social capital?" *Social Science & Medicine* 61(6): 1159–1173. https://doi.org/10.1016/j.socscimed.2005.01.017; Wood, L. J., B. Giles-Corti, M. K. Bulsara and D. A. Bosch (2007). "More than a furry companion: The ripple effect of companion animals on neighborhood interactions and sense of community." *Society & Animals* 15(1): 43–56. https://doi.org/10.1163/156853007X169333; Wood, L., K. Martin, H. Christian, A. Nathan, C. Lauritsen, S. Houghton, I. Kawachi and S. McCune (2015). "The Pet Factor – Companion Animals as a Conduit for Getting to Know People, Friendship Formation and Social Support." *PLOS ONE* 10(4). https://doi.org/10.1371/journal.pone.0122085

23 Gee, N. R., E. N. Crist and D. N. Carr (2010). "Preschool Children Require Fewer Instructional Prompts to Perform a Memory Task in the Presence of a Dog." *Anthrozoös* 23(2): 173–184. https://doi.org/10.2752/175303710X12682332910051
Gee, N. R., J. K. Gould, C. C. Swanson and A. K. Wagner (2012). "Preschoolers Categorize Animate Objects Better in the Presence of a Dog." *Anthrozoös* 25(2): 187–198. https://doi.org/10.2752/175303712X13316289505387

24 Gee, N., E. Friedmann, V. Coglitore, A. Fisk and M. Stendahl (2015). "Does Physical Contact with a Dog or Person Affect Performance of a Working Memory Task?" *Anthrozoös* 28: 483–500. https://doi.org/10.1080/08927936.2015.1052282

25 Purewal, R., R. Christley, K. Kordas, C. Joinson, K. Meints, N. Gee and C. Westgarth (2017). "Companion Animals and Child/Adolescent Development: A Systematic Review of the Evidence." *International Journal of Environmental Research and Public Health* 14(3): 234. https://doi.org/10.3390/ijerph14030234

26 Miles, J. N. V., L. Parast, S. H. Babey, B. A. Griffin and J. M. Saunders (2017). "A Propensity-Score-Weighted Population-Based Study of the Health Benefits of Dogs and Cats for Children." *Anthrozoös* 30(3): 429–440. https://doi.org/10.1080/08927936.2017.1335103

27 Gadomski, A. M., M. B. Scribani, N. Krupa, P. Jenkins, Z. Nagykaldi and A. L. Olson (2015). "Pet Dogs and Children's Health: Opportunities for Chronic Disease Prevention?" *Preventing Chronic Disease* 12: E205. http://dx.doi.org/10.5888/pcd12.150204

28 Tower, R. B. and M. Nokota (2006). "Pet companionship and depression: results from a United States Internet sample." *Anthrozoös* 19(1): 50–64. https://doi.org/10.2752/089279306785593874

29 Diener, E. (2000). "Subjective well-being: The science of happiness and a proposal for a national index." *American Psychologist* 55(1): 34–43. https://www.semanticscholar.org/paper/Subjective-well-being.-The-science-of-happiness-and-Diener/1b2a60c638bb5ac8b982c2ece09140f971c8c608; Emmerling, J. and S. Qari (2017). "Car ownership and hedonic adaptation." *Journal of Economic Psychology* 61: 29–38, https://doi.org/10.1016/j.joep.2017.02.014

30 Cohen, S. and T. A. Wills (1985). "Stress, social support, and the buffering hypothesis." *Psychological Bulletin* 98(2): 310–357. https://doi.org/10.1037/0033-2909.98.2.310

CHAPTER 6

1 For a deeper discussion of littermate syndrome and whether it exists, see Fratt, K. (2019) "There's no scientific reason to believe littermate syndrome exists" *The IAABC Journal*. https://winter2019.iaabcjournal.org/littermate-syndrome (Accessed 1 Feb 2021)

2 MacLean, E. L., N. Snyder-Mackler, B. M. vonHoldt and J. A. Serpell (2019). "Highly heritable and functionally relevant breed differences in dog behaviour." *Proceedings of the Royal Society B: Biological Sciences* 286(1912): 20190716. https://doi.org/10.1098/rspb.2019.0716

3 Fadel, F. R., P. Driscoll, M. Pilot, H. Wright, H. Zulch and D. Mills (2016). "Differences in Trait Impulsivity Indicate Diversification of Dog Breeds into Working and Show Lines." *Scientific Reports* 6(1): 22162. https://doi.org/10.1038/srep22162

4 Podberscek, A. L. and J. A. Serpell (1997). "Environmental influences on the expression of aggressive behaviour in English Cocker Spaniels." *Applied Animal Behaviour Science* 52(3/4): 215–227. https://doi.org/10.1016/S0168-1591(96)01124-0

Strandberg, E., J. Jacobsson and P. Saetre (2005). "Direct genetic, maternal and litter effects on behaviour in German shepherd dogs in Sweden." *Livestock Production Science* 93(1): 33–42. https://doi.org/10.1016/j.livprodsci.2004.11.004

Pérez-Guisado, J., R. Lopez-Rodríguez and A. Muñoz-Serrano (2006). "Heritability of dominant-aggressive behaviour in English Cocker Spaniels." *Applied Animal Behaviour Science* 100(3/4): 219–227. https://doi.org/10.1016/j.applanim.2005.11.005

Saetre, P., E. Strandberg, P. E. Sundgren, U. Pettersson, E. Jazin and T. F. Bergström (2006). "The genetic contribution to canine personality." *Genes Brain, and Behavior* 5(3): 240–248. https://doi.org/10.1111/j.1601-183X.2005.00155.x

Liinamo, A. E., L. v. d. Berg, P. A. J. Leegwater, M. B. H. Schilder, J. A. M. v. Arendonk and B. A. v. Oost (2007). "Genetic variation in aggression-related traits in Golden Retriever dogs." *Applied Animal Behaviour Science* 104(1/2): 95–106. https://doi.org/10.1016/j.applanim.2006.04.025

5 Westgarth, C., K. Reevell, and R. Barclay (2012). "Association between prospective owner viewing of the parents of a puppy and later referral for behavioural problems." *Veterinary Record* 170(19). https://doi.org/10.1136/vr.100138

6 Graham, T. M., K. J. Milaney, C. L. Adams and M. J. Rock (2018). 'Pets Negotiable': How Do the Perspectives of Landlords and Property Managers Compare with Those of Younger Tenants with Dogs?" *Animals* 8(3): 13. https://doi.org/10.3390/ani8030032.

7 Norman, C., J. Stavisky and C. Westgarth (2020). "Importing rescue dogs into the UK: reasons, methods and welfare considerations." Vet Rec 186(8): 248, https://doi.org/10.1136/vr.105380

8 Griffin, K. E., E. John, T. Pike and D. S. Mills (2020). "Can This Dog Be Rehomed to You? A Qualitative Analysis and Assessment of the Scientific Quality of the Potential Adopter Screening Policies and Procedures of Rehoming Organisations." *Frontiers in Veterinary Science* 7(1121). https://doi.org/10.3389/fvets.2020.617525

9 Howell, T. J., T. King and P. C. Bennett (2015). "Puppy parties and beyond: the role of early age socialization practices on adult dog behavior." *Veterinary Medicine* (Auckland, N.Z.) 6: 143–153. https://doi.org/10.2147/VMRR.S62081
Dietz, L., A.-M. K. Arnold, V. C. Goerlich-Jansson and C. M. Vinke (2018). "The importance of early life experiences for the development of behavioural disorders in domestic dogs." *Behaviour* 155(2–3): 83. https://doi.org/10.1163/1568539X-00003486

10 Scott, J. P. and J. L. Fuller (1965). *Genetics and Social Behavior of the Dog.* University of Chicago Press

11 Wauthier, L. M., J. M. Williams and Scottish Society for the Prevention of Cruelty to Animals (2018). "Using the mini C-BARQ to investigate the effects of puppy farming on dog behaviour." *Applied Animal Behaviour Science* 206: 75–86. https://doi.org/10.1016/j.applanim.2018.05.024
McMillan, F. D., J. A. Serpell, D. L. Duffy, E. Masaoud and I. R. Dohoo (2013). "Differences in behavioral characteristics between dogs obtained as puppies from pet stores and those obtained from noncommercial breeders." *Journal of the American Veterinary Medical Association* 242(10): 1359–1363. https://doi.org/10.2460/javma.242.10.1359

12 The UK government has a "Petfished" campaign to try to prevent these practices from fooling people (https://getyourpetsafely.campaign.gov.uk) and is also introducing "Lucy's Law" to ban third-party sales of puppies or kittens: https://www.gov.uk/government/news/lucys-law-spells-the-beginning-of-the-end-for-puppy-farming (Accessed 1 Feb 2021)

13 Brie's full name is actually Brienne of Tarth, because when we got her she had such a serious, worried look on her face all of the time. If you know, you know.

14 McMillan, F., D. Duffy and J. Serpell (2011). "Mental health of dogs formerly used as 'breeding stock' in commercial breeding establishments." *Applied Animal Behaviour Science* 135 (1), 86. https://doi.org/10.1016/j.applanim.2011.09.006

15 The Puppy Contract: https://puppycontract.org.uk (Accessed 1 Feb 2021)

16 Examples include Puppy Culture https://shoppuppyculture.com and The UK Kennel Club Guide "Rearing and raising your puppies" https://www.thekennelclub.org.uk/dog-breeding/first-time-breeders/rearing-and-raising-your-puppies (Accessed 1 Feb 2021)

17 Developed by Claire Stewart.

18 Yin, S. (2015) Checklist for Socialisation. https://drsophiayin.com/app/uploads/2015/12/Socialization_Checklist.pdf (Accessed 1 Feb 2021)

19 Provided by Raychel Hill.

20 Serpell, J. A. and D. L. Duffy (2016). "Aspects of Juvenile and Adolescent Environment Predict Aggression and Fear in 12-Month-Old Guide Dogs." *Frontiers in Veterinary Science* 3(49). https://doi.org/10.3389/fvets.2016.00049

21 Westgarth, C. and F. Watkins (2015). "A qualitative investigation of the perceptions of female dog-bite victims and implications for the prevention of dog bites." *Journal of Veterinary Behavior: Clinical Applications and Research* 10(6): 479–488. https://doi.org/10.1016/j.jveb.2015.07.035

22 Hemenway, D. (2013). "Three common beliefs that are impediments to injury prevention." *Injury Prevention* 19(4. http://dx.doi.org/10.1136/injuryprev-2012-040507

23 Owczarczak-Garstecka, S. C., F. Watkins, R. Christley and C. Westgarth (2018). "Online videos indicate human and dog behaviour preceding dog bites and the context in which bites occur." *Scientific Reports* 8(1): 7147. https://doi.org/10.1038/s41598-018-25671-7

24 Oxley, J. A., R. Christley and C. Westgarth (2018). "Contexts and consequences of dog bite incidents." *Journal of Veterinary Behavior: Clinical Applications and Research* 23(Supplement C): 33–39. https://doi.org/10.1016/j.jveb.2017.10.005

25 Newman, J., R. M. Christley, C. Westgarth and K. M. Morgan (2017). "Chapter 10: Risk Factors for Dog Bites – An Epidemiological Perspective", in Mills, D S. and C. Westgarth, *Dog Bites: A Multidisciplinary Perspective.* Sheffield: 5M Publishing, pp. 133–158

26 One interesting dataset on this is the CBARQ online survey of dog behavioural characteristics as reported by owners. The findings seem plausible; however, there was no adjustment for potential confounding factors such as the type of people who own particular breeds and how they interact with/manage their dogs. Duffy, D. L., Y. Y. Hsu and J. A. Serpell (2008). "Breed differences in canine aggression." *Applied Animal Behaviour Science* 114(3-4): 441–460. https://doi.org/10.1016/j.applanim.2008.04.006

27 Interestingly, Shih Tzus were found to be the third most common breed causing serious dogs bites to children in Liverpool around the same time that a study of mine showed that it was also the second most common breed children in Liverpool lived with – the first, and thus unsurprisingly also the cause of the most bites, being the Staffordshire bull terrier. The authors erroneously concluded that the Staffordshire bull terrier is the breed most likely to bite, because they compared their numbers with UK-wide statistics for breeds, which does not accurately reflect which dog breeds children in Liverpool live with.

28 Cornelissen, J. M. R. and H. Hopster (2010). "Dog bites in The Netherlands: A study of victims, injuries, circumstances and aggressors to support evaluation of breed specific legislation." *Veterinary Journal* 186(3): 292–298. https://doi.org/10.1016/j.tvjl.2009.10.001

Nilson, F., J. Damsager, J. Lauritsen and C. Bonander (2018). "The effect of breed-specific dog legislation on hospital treated dog bites in Odense, Denmark: A time series intervention study." *PLOS ONE* 13(12): 8. https://doi.org/10.1371/journal.pone.0208393

Mora, E., G. M. Fonseca, P. Navarro, A. Castano and J. Lucena (2018). "Fatal dog attacks in Spain under a breed-specific legislation: A ten-year retrospective study." *Journal of Veterinary Behavior: Clinical Applications and Research* 25: 76–84. https://doi.org/10.1016/j.jveb.2018.03.011

Creedon, N. and P. S. O. Suilleabhain (2017). "Dog bite injuries to humans and the use of breed-specific legislation: a comparison of bites from legislated and non-legislated dog breeds." *Irish Veterinary Journal* 70(1): 23. https://doi.org/10.1186/s13620-017-0101-1

29 Gov.uk. "Controlling your dog in public." https://www.gov.uk/control-dog-public (Accessed 2 Feb 2021)

30 Hence why the Dogs Trust educational programmes in schools begin at age seven; Baatz, A., K. L. Anderson, R. Casey, M. Kyle, K. M. McMillan, M. Upjohn and H. Sevenoaks (2020). "Education as a tool for improving canine welfare: Evaluating the effect of an education workshop on attitudes to responsible dog ownership and canine welfare in a sample of Key Stage 2 children in the United Kingdom." *PLOS ONE* 15(4): e0230832. https://doi.org/10.1371/journal.pone.0230832

31 Westgarth, C., M. Brooke and R. M. Christley (2018). "How many people have been bitten by dogs? A cross-sectional survey of prevalence, incidence and factors associated with dog bites in a UK community." *Journal of Epidemiology and Community Health* 72: 331–336. http://dx.doi.org/10.1136/jech-2017-209330

CHAPTER 7

1 Wolf biologist David Mech speaks here https://www.youtube.com/watch?v=tNtFgdwTsbU

2 Bradshaw, John (2011). *In Defence of Dogs*. London: Penguin

3 For more of my thoughts on the dominance in dogs debate, see: Westgarth, C. (2016). "Why nobody will ever agree about dominance in dogs." *Journal of Veterinary Behavior: Clinical Applications and Research* 11: 99–101. https://doi.org/10.1016/j.jveb.2015.02.004

4 China, L., D. S. Mills and J. J. Cooper (2020). "Efficacy of Dog Training With and Without Remote Electronic Collars vs. a Focus on Positive Reinforcement." *Frontiers in Veterinary Science* 7(508). https://doi.org/10.3389/fvets.2020.00508

5 Hiby, E. F., N. J. Rooney and J. W. S. Bradshaw (2004). "Dog training methods: their use, effectiveness and interaction with behaviour and welfare." *Animal Welfare* 13(1): 63–69. https://www.ingentaconnect.com/content/ufaw/aw/2004/00000013/00000001/art00010

Dodman, N. H., D. C. Brown and J. A. Serpell (2018). "Associations between owner personality and psychological status and the prevalence of canine behavior problems." *PLOS ONE* 13(2): e0192846. https://doi.org/10.1371/journal.pone.0192846

LaFollette, M. R., K. E. Rodriguez, N. Ogata and M. E. O'Haire (2019). "Military Veterans and Their PTSD Service Dogs: Associations Between Training Methods, PTSD Severity, Dog Behavior, and the Human–Animal Bond." *Frontiers in Veterinary Science* 6: 11. https://doi.org/10.3389/fvets.2019.00023

6 Vieira de Castro, A. C., D. Fuchs, G. M. Morello, S. Pastur, L. de Sousa and I. A. S. Olsson (2020). "Does training method matter? Evidence for the negative impact of aversive-based methods on companion dog welfare." PLOS ONE 15(12): e0225023. https://doi.org/10.1371/journal.pone.0225023

7 Welfare in Dog Training. https://www.dogwelfarecampaign.org

8 Feuerbacher, E. N. and C. D. Wynne (2014). "Most domestic dogs (*Canis lupus familiaris*) prefer food to petting: population, context, and schedule effects in concurrent choice." *Journal of the Experimental Analysis of Behavior* 101(3): 385–405. https://doi.org/10.1002/jeab.81

CHAPTER 8

1 Westgarth, C., R. M. Christley, G. Marvin and E. Perkins (2020). "Functional and recreational dog walking practices in the UK." *Health Promotion International*. https://doi.org/10.1093/heapro/daaa051

2 https://dogtime.com/dog-health/general/1530-dog-training-walking-on-leash-dunbar

3 Carter, A., D. McNally and A. Roshier (2020). "Canine collars: an investigation of collar type and the forces applied to a simulated neck model." *Veterinary Record* 187(7): e52–e52. https://doi.org/10.1136/vr.105681
Summarized at https://www.companionanimalpsychology.com/2020/05/flat-collars-risk-damage-to-dogs-necks.html?fbclid=IwAR0T3xtszl2NIKgc8JCtP5FiqywohR_kOgq_SrY114bHV3DNGIqIs1MN6R0

CHAPTER 9

1 Rhodes, R. E., H. Murray, V. A. Temple, H. Tuokko and J. W. Higgins (2012). "Pilot study of a dog walking randomized intervention: Effects of a focus on canine exercise." *Preventive Medicine* 54(5): 309–312. https://doi.org/10.1016/j.ypmed.2012.02.014

2 Rhodes, R. E., M. Baranova, H. Christian and C. Westgarth (2020). "Increasing physical activity by four legs rather than two: systematic review of dog-facilitated physical activity interventions." *British Journal of Sports Medicine* 54: 1202–1207. http://dx.doi.org/10.1136/bjsports-2019-101156

3 Westgarth, C., R. M. Christley, G. Marvin and E. Perkins (2017). "I Walk My Dog Because It Makes Me Happy: A Qualitative Study to Understand Why Dogs Motivate Walking and Improved Health." *International Journal of Environmental Research and Public Health* 14(8). https://doi.org/10.3390/ijerph14080936

4 Westgarth, C., R. M. Christley, G. Marvin and E. Perkins (2019). "The Responsible Dog Owner: The Construction of Responsibility." *Anthrozoös* 32(5): 631–646. https://doi.org/10.1080/08927936.2019.1645506

5 Westgarth, C., R. M. Christley and H. E. Christian (2014). "How might we increase physical activity through dog walking?: A comprehensive review of dog walking correlates." *International Journal of Behavioral Nutrition and Physical Activity* 11 (83). https://doi.org/10.1186/1479-5868-11-83

6 Brown, S. G. and R. E. Rhodes (2006). "Relationships among dog ownership and leisure-time walking in western Canadian adults." *American Journal of Preventive Medicine* 30(2): 131–136. https://doi.org/10.1016/j.amepre.2005.10.007

Cutt, H., B. Giles-Corti and M. Knuiman (2008). "Encouraging physical activity through dog walking: Why don't some owners walk with their dog?" *Preventive Medicine* 46(2): 120–126. https://doi.org/10.1016/j.ypmed.2007.08.015

7 Lim, C. and R. E. Rhodes (2016). "Sizing up physical activity: The relationships between dog characteristics, dog owners' motivations, and dog walking." *Psychology of Sport and Exercise* 24: 65–71. https://doi.org/10.1016/j.psychsport.2016.01.004

8 Pickup, E., A. J. German, E. Blackwell, M. Evans and C. Westgarth (2017). "Variation in activity levels amongst dogs of different breeds: results of a large online survey of dog owners from the UK." *Journal of Nutritional Science* 6(e10), https://doi.org/10.1017/jns.2017.7

9 The Kennel Club. "Puppy and dog walking tips." https://www.thekennelclub.org.uk/dog-training/getting-started-in-dog-training/dog-training-and-games/puppy-and-dog-walking-tips/#:~:text=A%20good%20rule%20of%20thumb,go%20out%20for%20much%20longer (Accessed 4 Feb 2021)

10 Westgarth, C., M. Knuiman and H. E. Christian (2016). "Understanding how dogs encourage and motivate walking: cross-sectional findings from RESIDE." *BMC Public Health* 16(1): 1019. https://doi.org/10.1186/s12889-016-3660-2.

11 Christian, H., B. Giles-Corti and M. Knuiman (2010). "'I'm Just a'-Walking the Dog': Correlates of Regular Dog Walking." *Family & Community Health* 33(1): 44–52. doi: 10.1097/FCH.0b013e3181c4e208

12 Westgarth, C., R. M. Christley, G. Marvin and E. Perkins (2020). "Functional and recreational dog walking practices in the UK." *Health Promotion International*. https://doi.org/10.1093/heapro/daaa051

13 For a seminal paper in this field, see Ajzen, I. (1991). "The theory of planned behaviour." *Organisational Behaviour and Human Performance* 50: 179–211. https://doi.org/10.1016/0749-5978(91)90020-T

14 Brown, S. G. and R. E. Rhodes (2006). "Relationships among dog ownership and leisure-time walking in western Canadian adults." *American Journal of Preventive Medicine* 30(2): 131–136. 10.1016/j.amepre.2005.10.007

15 Rhodes, R. E. and C. Lim (2016). "Understanding action control of daily walking behavior among dog owners: a community survey." *BMC Public Health* 16(1). 1165, https://doi.org/10.1186/s12889-016-3814-2

16 Rhodes, R. E. (2014). "Bridging the physical activity intention-behaviour gap: contemporary strategies for the clinician." *Applied Physiology Nutrition and Metabolism* 39(1): 105–107. https://doi.org/10.1139/apnm-2013-0166; Rhodes, R. E., I. Janssen, S. S. D. Bredin, D. E. R. Warburton and A. Bauman (2017). "Physical activity: Health impact, prevalence, correlates and interventions." *Psychology and Health*, 32(8): 942–975. https://doi.org/10.1080/08870446.2017.1325486

17 Wu, Y.-T., R. Luben and A. Jones (2017). "Dog ownership supports the maintenance of physical activity during poor weather in older English adults: cross-sectional results from the EPIC Norfolk cohort." *Journal of Epidemiology and Community Health* **71**(9): 905–911. http://dx.doi.org/10.1136/jech-2017-208987

18 Westgarth, C., R. M. Christley, G. Marvin and E. Perkins (2020). "Functional and recreational dog walking practices in the UK." *Health Promotion International*. https://doi.org/10.1093/heapro/daaa051

19 Belshaw, Z., R. Dean and L. Acher (2020). "Slower, shorter, sadder: a qualitative study exploring how dog walks change when the canine participant develops osteoarthritis." *BMC Veterinary Research* **16**(1). https://doi.org/10.1186/s12917-020-02293-8

20 Brown, K. and R. Dilley (2012). "Ways of knowing for 'response-ability' in more-than-human encounters: the role of anticipatory knowledges in outdoor access with dogs." *Area* **44**(1): 37–45. https://www.jstor.org/stable/41406043

CHAPTER 11

1 For a fantastic example of what can be achieved through backchaining, do an internet search for "mouse agility".

CHAPTER 12

1 Rodriguez, K. E., J. Bibbo, S. Verdon and M. E. O'Haire (2020). "Mobility and medical service dogs: a qualitative analysis of expectations and experiences." *Disability and Rehabilitation-Assistive Technology* **15**(5): 499–509. https://doi.org/10.1080/17483107.2019.1587015

2 O'Haire, M. E. and K. E. Rodriguez (2018). "Preliminary efficacy of service dogs as a complementary treatment for posttraumatic stress disorder in military members and veterans." *Journal of Consulting and Clinical Psychology* **86**(2): 179–188. https://doi.org/10.1037/ccp0000267

3 Thirty per cent of the analyses showed a positive effect but 68 per cent of the analyses didn't show any evidence of an effect. Rodriguez, K. E., J. Bibbo and M. E. O'Haire (2020). "The effects of service dogs on psychosocial health and wellbeing for individuals with physical disabilities or chronic conditions." *Disability and Rehabilitation* **42**(10): 1350–1358. https://doi.org/10.1080/09638288.2018.1524520

4 Rodriguez, K. E., J. Greer, J. K. Yatcilla, A. M. Beck and M. E. O'Haire (2020). "The effects of assistance dogs on psychosocial health and wellbeing: A systematic literature review." *PLOS ONE* **15**(12): e0243302. https://doi.org/10.1371/journal.pone.0243302

5 Bremhorst, A., P. Mongillo, T. Howell and L. Marinelli (2018). "Spotlight on Assistance Dogs – Legislation, Welfare and Research." *Animals* **8**(8): 129. https://doi.org/10.3390/ani8080129

6 Gravrok, J., D. Bendrups, T. Howell and P. Bennett (2019). "Beyond the Benefits of Assistance Dogs: Exploring Challenges Experienced by First-Time Handlers." *Animals* **9**(5): 203. https://doi.org/10.3390/ani9050203

7 Unfortunately, some unscrupulous organizations are trying to make money out of training assistance dogs for the highest bidder, and are ruining vulnerable people's lives by supplying dogs that do not do what was promised – or are even sick. For more information about recognized, accredited assistance-dog organizations that work to a high standard, see Assistance Dogs International https://assistancedogsinternational.org and if you are in the UK, Assistance Dogs UK https://www.assistancedogs.org.uk

CHAPTER 13

1 Asher, L., G. C. W. England, R. Sommerville and N. D. Harvey (2020). "Teenage dogs? Evidence for adolescent-phase conflict behaviour and an association between attachment to humans and pubertal timing in the domestic dog." *Biology Letters* 16(5): 20200097. https://doi.org/10.1098/rsbl.2020.0097

2 Mills, D. S., I. Demontigny-Bédard, M. Gruen, M. P. Klinck, K. J. McPeake, A. M. Barcelos, L. Hewison, H. Van Haevermaet, S. Denenberg, H. Hauser, C. Koch, K. Ballantyne, C. Wilson, C. V. Mathkari, J. Pounder, E. Garcia, P. Darder, J. Fatjó and E. Levine (2020). "Pain and Problem Behavior in Cats and Dogs." *Animals* 10(2): 318. https://doi.org/10.3390/ani10020318

3 In the UK I would recommend members of the Association of Pet Behaviour Counsellors (www.apbc.org.uk) or those eligible to be listed by the Animal Behaviour and Training Council (www.abtc.org.uk) as "Clinical Animal Behaviourist" or "Veterinary Behaviourist" (which includes members of the APBC).

CHAPTER 14

1 Dogs Trust (2020). "The impact of COVID-19 lockdown restrictions on dogs and dog owners in the UK." https://www.dogstrust.org.uk/help-advice/research/research-papers/201020_covid%20report_v8.pdf; Christley, R. M., J. K. Murray, K. L. Anderson, E. L. Buckland, R. A. Casey, N. D. Harvey, L. Harris, K. E. Holland, K. M. McMillan, R. Mead, S. C. Owczarczak-Garstecka and M. M. Upjohn (2021). "Impact of the First COVID-19 Lockdown on Management of Pet Dogs in the UK." *Animals* 11(1): 5. https://doi.org/10.3390/ani11010005; Holland, K. E., S. C. Owczarczak-Garstecka, K. L. Anderson, R. A. Casey, R. M. Christley, L. Harris, K. M. McMillan, R. Mead, J. K. Murray, L. Samet and M. M. Upjohn (2021). "'More Attention than Usual': A Thematic Analysis of Dog Ownership Experiences in the UK during the First COVID-19 Lockdown." *Animals* 11(1): 240. https://doi.org/10.3390/ani11010240

2 Daly, M., A. R. Sutin and E. Robinson (2020). "Longitudinal changes in mental health and the COVID-19 pandemic: evidence from the UK Household Longitudinal Study." *Psychological Medicine*: 1–10. https://doi.org/10.1017/S0033291720004432

3 Ratschen, E., E. Shoesmith, L. Shahab, K. Silva, D. Kale, P. Toner, C. Reeve and D. S. Mills (2020). "Human–animal relationships and interactions during the Covid-19 lockdown phase in the UK: Investigating links with

mental health and loneliness." *PLOS ONE* **15**(9): e0239397. https://doi.org/10.1371/journal.pone.0239397

4 Oliva, J. L. and K. L. Johnston (2020). "Puppy love in the time of Corona: Dog ownership protects against loneliness for those living alone during the COVID-19 lockdown." *International Journal of Social Psychiatry*: 11. https://doi.org/10.1177/0020764020944195

5 Applebaum, J. W., C. A. Tomlinson, A. Matijczak, S. E. McDonald and B. A. Zsembik (2020). "The Concerns, Difficulties, and Stressors of Caring for Pets during COVID-19: Results from a Large Survey of US Pet Owners." *Animals* **10**(10): 14. https://doi.org/10.3390/ani10101882

6 Bowen, J., E. García, P. Darder, J. Argüelles and J. Fatjó (2020). "The effects of the Spanish COVID-19 lockdown on people, their pets and the human-animal bond." *Journal of Veterinary Behavior* **40**: 75–91. https://doi.org/10.1016/j.jveb.2020.05.013

7 Morgan, L., A. Protopopova, R. I. D. Birkler, B. Itin-Shwartz, G. A. Sutton, A. Gamliel, B. Yakobson and T. Raz (2020). "Human–dog relationships during the COVID-19 pandemic: booming dog adoption during social isolation." *Humanities and Social Sciences Communications* **7**(1): 155. https://doi.org/10.1057/s41599-020-00649-x; Four Paws UK (2020). "The Impact of COVID-19 on the Online Puppy Trade: United Kingdom." https://media.4-paws.org/a/b/9/e/ab9ebce5a46b5a2ba4935de29573b199fc83c9b1/Covid19%20Puppy%20Sales%20UK%20Results%20June%202020.pdf (Accessed 19 Jan 2021)

8 BBC News, Watson, C. (2020). "Puppy prices soar during coronavirus lockdown." https://www.bbc.co.uk/news/uk-scotland-54115646 (Accessed 19 Jan 2021)

9 Dogs Trust (2020). "Huge price hikes for popular breeds during lockdown revealed." https://www.dogstrust.org.uk/news-events/news/2020/huge-price-hikes-for-popular-breeds-during-lockdown-revealed (Accessed 19 Jan 2021)

10 The Kennel Club (2020). "The Covid-19 puppy boom – one in four admit impulse buying a pandemic puppy." https://www.thekennelclub.org.uk/media-centre/2020/august/the-covid-19-puppy-boom-one-in-four-admit-impulse-buying-a-pandemic-puppy (Accessed 19 Jan 2021)

11 Kent Smith, E. (2021). "Lockdown pups sent packing by families with no time for walkies." *The Times*, London. https://www.thetimes.co.uk/article/lockdown-pups-sent-packing-by-families-with-no-time-for-walkies-8h68htstb (Accessed 19 Jan 2021)

12 Association of Pet Behaviour Counsellors (2020). "COVID-19 Information." https://www.apbc.org.uk/covid-19-information (Accessed 19 Jan 2021); Dogs Trust (2020). "Lockdown Advice: Introducing your puppy to the world around them." https://www.dogstrust.org.uk/help-advice/behaviour/puppy-socialisation-introduction (Accessed 19 Jan 2021)

13 Grice, H. (2020). "Puppy socialisation bingo during Covid-19." https://www.doglistener.tv/wp-content/uploads/puppy-socialisation-bingoCR.pdf?fbclid=IwAR3-R_Txz-KCJ-3lVd-6_R5RpvM3xMGvKzAx1T6kF2wHIKtHOfiEB4ae9Go (Accessed 25 Jan 2021)

14 Brackenridge, S., L. K. Zottarelli, E. Rider and B. Carlsen-Landy (2012). "Dimensions of the Human–Animal Bond and Evacuation Decisions among

Pet Owners during Hurricane Ike." *Anthrozoös* 25(2): 229–238. https://doi.org/10.2752/175303712X13316289505503; Chadwin, R. (2017). "Evacuation of Pets During Disasters: A Public Health Intervention to Increase Resilience." *American Journal of Public Health.* 107(9): 1413–1417 https://doi.org/10.2105/AJPH.2017.303877; Taylor, M., E. Lynch, P. Burns and G. Eustace (2015). "The preparedness and evacuation behaviour of pet owners in emergencies and natural disasters." *Australian Emergency Management Institute* https://knowledge.aidr.org.au/resources/ajem-apr-2015-the-preparedness-and-evacuation-behaviour-of-pet-owners-in-emergencies-and-natural-disasters (Accessed 19 Jan 2021)

15 Whittaker, J., M. Taylor and C. Bearman (2020). "Why don't bushfire warnings work as intended? Responses to official warnings during bushfires in New South Wales, Australia." *International Journal of Disaster Risk Reduction* 45: 10. https://doi.org/10.1016/j.ijdrr.2020.101476; Thompson, K. R., L. Haigh and B. P. Smith (2018). "Planned and ultimate actions of horse owners facing a bushfire threat: Implications for natural disaster preparedness and survivability." *International Journal of Disaster Risk Reduction* 27: 490–498. https://doi.org/10.1016/j.ijdrr.2017.11.013

16 Tanaka, A., J. Saeki, S. Hayama and P. H. Kass (2019). "Effect of Pets on Human Behavior and Stress in Disaster." *Frontiers in Veterinary Science* 6: 8. https://doi.org/10.3389/fvets.2019.00113

17 Applebaum, J. W., B. L. Adams, M. N. Eliasson, B. A. Zsembik and S. E. McDonald (2020). "How pets factor into healthcare decisions for COVID-19: A One Health perspective." *One Health* 11: 100176. https://doi.org/10.1016/j.onehlt.2020.100176; also summarized by Hal Herzog: Herzog, H. (no date) "Why Do People Risk Their Own Health for Their Pets?" *Psychology Today*, 24 October. https://www.psychologytoday.com/us/blog/animals-and-us/202010/why-do-people-risk-their-own-health-their-pets (Accessed 8 Feb 2021)

18 Onukem, M. (2016). "Assessment of emergency/disaster preparedness and awareness for animal owners in Canada." *International Journal of Emergency Services* 5(2): 212–222. https://doi.org/10.1108/IJES-07-2016-0012

CHAPTER 15

1 Szabó, D., N. R. Gee and Á. Miklósi (2016). "Natural or pathologic? Discrepancies in the study of behavioral and cognitive signs in aging family dogs." *Journal of Veterinary Behavior* 11: 86–98. https://doi.org/10.1016/j.jveb.2015.08.003; Wallis, L. J., D. Szabó, B. Erdélyi-Belle and E. Kubinyi (2018). "Demographic Change Across the Lifespan of Pet Dogs and Their Impact on Health Status." *Frontiers in Veterinary Science* 5: 200. https://doi.org/10.3389/fvets.2018.00200

2 Kraus, C., S. Pavard and D. E. Promislow (2013). "The Size–Life Span Trade-Off Decomposed: Why Large Dogs Die Young." *The American Naturalist* 181(4): 492–505. https://doi.org/10.1086/669665

3 Pet Health Network. "How old is your pet in people years?" https://idexxcom-live-b02da1e51e754c9cb292133b-9c56c33.aldryn-media.com/

filer_public/a3/20/a320508f-3970-4938-956f-b79e0a887164/preventive-brochures-age-chart.pdf (Accessed 14 Jan 2021)

4 Chapagain, D., F. Range, L. Huber and Z. Virányi (2018). "Cognitive Aging in Dogs." *Gerontology* **64**(2): 165–171. https://doi.org/10.1159/000481621

5 Tufts University Cummings School of Veterinary Medicine (YEAR). "Comfort Diary for Dogs." https://yourfamilydogpodcast.com/wp-content/uploads/sites/46/2019/05/Tufts-Comfort-Diary-for-Dogs.pdf (Accessed 14 Jan 2021)

6 Find out about canine arthritis and how to spot it with Canine Arthritis Management's free expert veterinary advice. https://caninearthritis.co.uk/ (Accessed 14 Jan 2021)

7 Johnston, S. A. (1997). "Osteoarthritis: Joint Anatomy, Physiology, and Pathobiology." *Veterinary Clinics of North America: Small Animal Practice* 27(4): 699–723 https://doi.org/10.1016/S0195-5616(97)50076-3; Anderson, K. L., D. G. O'Neill, D. C. Brodbelt, D. B. Church, R. L. Meeson, D. Sargan and L. M. Collins (2018). "Prevalence, duration and risk factors for appendicular osteoarthritis in a UK dog population under primary veterinary care." *Scientific Reports*. 8(1): 1–12 https://doi.org/10.1038/s41598-018-23940-z

8 Kealy, R. D., D. F. Lawler, J. M. Ballam, S. L. Mantz, D. N. Biery, E. H. Greeley and Stowe, H. D. (2002). "Effects of diet restriction on life span and age-related changes in dogs." *Journal of the American Veterinary Medical Association* 220(9): 1315–1320. https://doi.org/10.2460/javma.2002.220.1315

9 Milgram, N. W., E. Head, S. C. Zicker, C. J. Ikeda-Douglas, H. Murphey, B. Muggenburg and C. W. Cotman (2005). "Learning ability in aged beagle dogs is preserved by behavioral enrichment and dietary fortification: a two-year longitudinal study." *Neurobiology of Aging* 26(1): 77–90. https://doi.org/10.1016/j.neurobiolaging.2004.02.014

10 Milgram, N. W., C. T. Siwak-Tapp, J. Araujo and E. Head (2006). "Neuroprotective effects of cognitive enrichment." *Ageing Research Reviews* 5(3): 354–369. https://doi.org/10.1016/j.arr.2006.04.004

11 Useful interactive lifestyle tool that highlights hazards for your dog both in the home and outside. https://caninearthritis.co.uk/lifestyle-tool (Accessed 14 Jan 2021)

12 Glickman, L. T., N. W. Glickman, G. E. Moore, E. M. Lund, G. C. Lantz and B. M. Pressler (2011). "Association between chronic azotemic kidney disease and the severity of periodontal disease in dogs." *Preventive Veterinary Medicine* 99(2–4): 193–200. https://doi.org/10.1016/j.prevetmed.2011.01.011

13 The Ohio State University Veterinary Medical Center (YEAR). "How do I know when it is time?" https://vet.osu.edu/vmc/sites/default/files/import/assets/pdf/hospital/companionAnimals/HonoringtheBond/HowDoIKnow When.pdf (Accessed 14 Jan 2021)

14 Blue Cross (no date). "Pet Loss." https://www.bluecross.org.uk/pet-loss (Accessed 14 Jan 2021)

15 In terms of anticipatory grief about dogs, there is the Ralph site "How to deal with anticipatory grief for a terminally ill or elderly pet." https://www.theralphsite.com/index.php?idPage=88 (Accessed 14 Jan 2021)

16 Viorst, J. (1987). *The Tenth Good Thing About Barney*. Upper Saddle River, NJ: Prentice Hall

CHAPTER 16

1 Mackenzie, J. S. and M. Jeggo (2019). "The One Health Approach: Why Is It So Important?" *Tropical Medicine and Infectious Disease* 4(2): 88. https://doi.org/10.3390/tropicalmed4020088; One Health Initiative. https://onehealthinitiative.com/mission-statement (Accessed 14 Jan 2021)

2 Beck, A. M. and A. H. Katcher (2003). "Future Directions in Human–Animal Bond Research." *American Behavioral Scientist* 47(1): 79–93. https://doi.org/10.1177/0002764203255214

3 Herzog, H. (no date). "Animals and Us: The psychology of human–animal interactions." *Psychology Today*. https://www.psychologytoday.com/gb/blog/animals-and-us (Accessed 14 Jan 2021.

4 Herzog, H. (2019). "The Sad Truth About Pet Ownership and Depression." *Psychology Today*. https://www.psychologytoday.com/gb/blog/animals-and-us/201912/the-sad-truth-about-pet-ownership-and-depression (Accessed 14 Jan 2021)

5 O'Haire, M. E. (2013). "Animal-Assisted Intervention for Autism Spectrum Disorder: A Systematic Literature Review." *Journal of Autism and Developmental Disorders* 43(7): 1606–1622. https://doi.org/10.1007/s10803-012-1707-5

6 Dolan, P., T. Peasgood and M. White (2008). "Do we really know what makes us happy? A review of the economic literature on the factors associated with subjective well-being." *Journal of Economic Psychology* 29(1): 94–122. https://doi.org/10.1016/j.joep.2007.09.001

INDEX